Understanding Stainless Steel

By

Alan Harrison

Published February 2009

By

British Stainless Steel Association

Broomgrove
59 Clarkehouse Road
Sheffield S10 2LE

Telephone: +44 (0) 114 267 1260
Fax: +44 (0) 114 266 1252
Email: enquiry@bssa.org.uk
Website: www.bssa.org.uk

ISBN 978-0-9561897-0-7

About the British Stainless Steel Association

The British Stainless Steel Association is a membership based organisation whose purpose is to promote the greater use of stainless steel in the UK and Ireland. Its strength lies in the breadth of its membership with companies from all sectors of the industry and all stages of the supply chain.

Originally formed as an association of stainless steel fabricators, the BSSA has steadily increased its scope of membership and range of activities over the years. In 2000, the BSSA was reformed and a small full-time staff was established to allow its services to be developed further. At the same time the Stainless Steel Advisory Service was launched which now operates as an integral part of the BSSA.

The BSSA has four principal areas of activity:

- Providing help and advice both through the Stainless Steel Advisory Service and via the website: **www.bssa.org.uk**;
- Training and education, including Starter Workshops, the Stainless Steel Specialist Course, open seminars and in-company bespoke workshops;
- Market development initiatives in sectors such as Architecture, Building and Construction, Water and Energy;
- Industry events, including forums, seminars, conferences and social functions.

The BSSA also provides benefits to its members through its involvement with other partner organisations such as the International Stainless Steel Forum (ISSF) and the European Stainless Steel Development Association (Euro Inox). It also works closely with other UK based organisations including UK Steel, the Steel Construction Institute, NAMTEC and the Metals Forum.

About the Author

Alan Harrison graduated in Metallurgy from Sheffield University in 1975. Since then he has been involved in the Sheffield steel industry. From 1975 to 1989 he was a metallurgist at British Steel River Don Works later Sheffield Forgemasters. His involvement in stainless steel began with a move in 1989 to British Steel Stainless later Avesta Sheffield, Avesta Polarit and Outokumpu Stainless. After excursions into market research and IT, he returned to the metallurgical world in 2006 on becoming the Technical Advisor to the BSSA.

Understanding Stainless Steel

The idea for this book grew out of a series of “Starter Workshops” run by the British Stainless Steel Association (BSSA). These 1-day seminars are designed for those who have little or no knowledge of stainless steel or who need a refresher in the basics. "Understanding Stainless Steel" complements the “Starter Workshops” and will help those involved in specifying, designing, buying, selling or fabricating this versatile product.

I hope that you will find that the book increases your knowledge of a material which is becoming increasingly important in the developed and developing world.

This book is dedicated to the thousands and men and women who have made the city of Sheffield a synonym for high quality steel. Although the “city of a thousand fires” has fewer steel companies than in its heyday it is still a significant producer of stainless and other high grade steels which underpins the whole economy. After all it was a Sheffielder who invented stainless steel.

I am grateful to my colleagues at the BSSA, particularly David Humphreys who was instrumental in developing the "Starter Workshops", and from BSSA Members who made helpful suggestions with the content of the book. A special thank you to my daughter, Joy, who read and commented on the initial draft.

Acknowledgments

I would particularly like to thank the following organisations for providing images, diagrams and assistance with the content of the Guide:

BSSA Marketing and Technical Committee
ArcelorMittal Stainless
Australian Stainless Steel Development Association
CRU
ELG Haniel
G-Tex Stainless
Jordan Manufacturing Ltd
Judith Duddle
Nickel Institute
Outokumpu Stainless
Pland Stainless
Rimex Metals
Stainless Restoration
Valbruna UK Ltd

Contents

Contents

Chapter 1 - The World of Stainless Steel

Our modern world would be unthinkable without stainless steel – or more accurately "steels". As we shall see, "stainless steel" covers a wide range of materials each suited to a particular set of conditions.

The following are a typical but by no means exhaustive list of applications:

Transport

Architecture

Civil Engineering

Domestic Appliances

Industrial Processing

Food and Beverage

Surgical and Medical

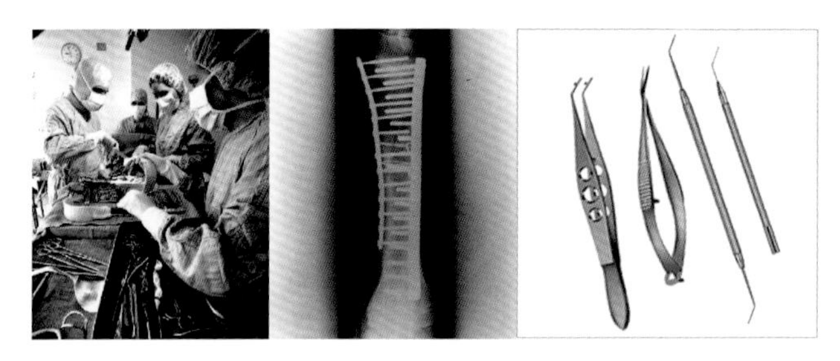

Street Furniture

Sculptures

Chapter 2 - A Brief History of Stainless Steel

In 2013, the world will be celebrating 100 years of stainless steel. This is the anniversary of its discovery in 1913 by one of Sheffield's pioneer metallurgists – Harry Brearley.

The story goes that he was trying various compositions of steel for gun barrels. One such batch of steel contained 13% Chromium. Apparently, he noticed that the batch of this steel had not gone rusty after several months. And so stainless steel was discovered.

However, the truth is probably more complicated than this and most western European countries and the US can find a similar hero, including:

Hans Goldschmidt
Germany

Development of carbon free chromium assisted the practical melting of stainless steels.

Albert Portevin
France

Studied a forerunner of type 430 stainless steel

Elwood Haynes
USA

Worked on a non-rusting razor steel.

As with most discoveries, a number of bits of knowledge came together at the right time. What is clear is that stainless steel took off quite significantly from the mid-1910s onwards.

Cutlery was the first major application and continues to consume large amounts of steel. But very quickly it was realised that stainless steel can be used for a wide range of applications.

As early as 1925, stainless steel was used in the chemical industry to store nitric acid. In 1929 the well known Chrysler Building was topped out with a stainless steel "crown". 80 years later this remains in good condition.

Since its invention, hundreds of applications have been added to the list. In recent times the growth in stainless steel consumption has outperformed other metallic materials as shown below:

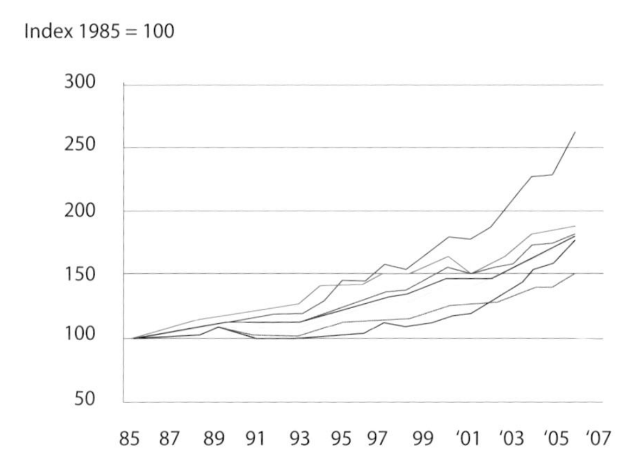

In the last 20 years, this growth was driven by two major factors:

- Reduction in cost of production: Over the years stainless steel has benefited from developments in the manufacturing processes which have led to cost reductions. Examples include the AOD vessel and continuous casting.

- General economic growth in the Pacific Rim region such as Korea and Taiwan: It is being continued in countries such as China and India. Stainless steel is used in consumer, infrastructure and industrial products, so all aspects of an economy support the growth in stainless steel.

It is forecast that such growth will continue in the existing industrial countries with the addition of Brazil, Russia and other Eastern European countries.

Chapter 3 - A Little Metallurgy

Basic Definitions

Most of these definitions would be scorned by a "proper" chemist or metallurgist but they will do for our purposes.

Atom – The basic building block of matter. Atoms are very small about 0.1 nanometres. This means that you could pack 10 million atoms on a line 1mm long.

Element – A chemical that contains only one sort of atom. Familiar elements are oxygen, silicon, iron, aluminium, sulphur, nitrogen.

Chemical Symbol – A shorthand way of denoting an element. Either one or two letters are used for each element. For example:

Fe = Iron (Fe from Latin ferrum)
Cr = Chromium
Mo = Molybdenum
Mn = Manganese
S = Sulphur
C = Carbon
Ni = Nickel
Ti = Titanium
Si = Silicon
N = Nitrogen

A fuller list can be found in chapter 15.

Metal – An element that is usually shiny, easy to form, conducts heat and electricity well. Iron, copper, aluminium, nickel, lead, zinc and chromium are metals. Metals normally exist as crystals or grains.

Crystal – A crystalline material is one in which the atoms are arranged in a regular 3 dimensional pattern. This does not mean that metals show a regular shape on a large scale like a quartz crystal.

Compound – A combination of two or more elements which forms a different material to any of the constituent elements. A well known compound is salt or sodium chloride. This is made from sodium which is a soft highly reactive metal and chlorine which is a poisonous green gas. Sodium chloride is a white, crystalline solid.

Molecule – The smallest part in which a compound can exist. A molecule of sodium chloride consists of one atom of sodium and one of chlorine.

Formula – A shorthand way of describing a compound. For example, NaCl is sodium chloride. Na is the symbol for sodium (sodium was also called natrium). Cl is the symbol for chlorine.

Alloy – Pure metals are not normally used for structural purposes as they are quite weak. Pure metals like copper and aluminium are used where thermal or electrical conductivity is the primary requirement. Alloys are combinations of a metal with other elements. The atoms of the different metals are found in the form of a solid solution or as chemical compounds. Each distinctive type of structure in an alloy is called a phase. Common examples of alloys are steel, brass, bronze and solders. Alloys are stronger than the constituent elements.

Steel – Steel is a common alloy. At its simplest it is an alloy of iron and carbon. Many different types of steel have been developed for different purposes. Additional elements in steel include silicon, manganese, chromium, nickel, molybdenum, vanadium, sulphur, phosphorus.

Heat Treatment – The properties of alloys can be affected by heating to and cooling from different temperatures. For steels, common heat treatments are:

> **Quenching –** A hardening process achieved by rapid cooling in water, oil or forced air. Steels in the as-quenched condition are too hard and brittle for practical use.
>
> **Tempering –** A treatment that softens a steel that has been hardened by quenching to give a practical combination of strength and toughness.

This basic information will help you to understand the technical content of everything from here onwards.

Chapter 4 - Why is Stainless Steel "Stainless"?

What Harry Brearley and others discovered was that chromium is the ingredient which gives stainless steels their "stainlessness". Look at the graph below. We can clearly see that increasing chromium (Cr) gives a much reduced metal loss in atmospheric conditions. When you get to 10.5% Cr, there is enough to define it as a stainless steel according to the current EN (European Norm) Standard.

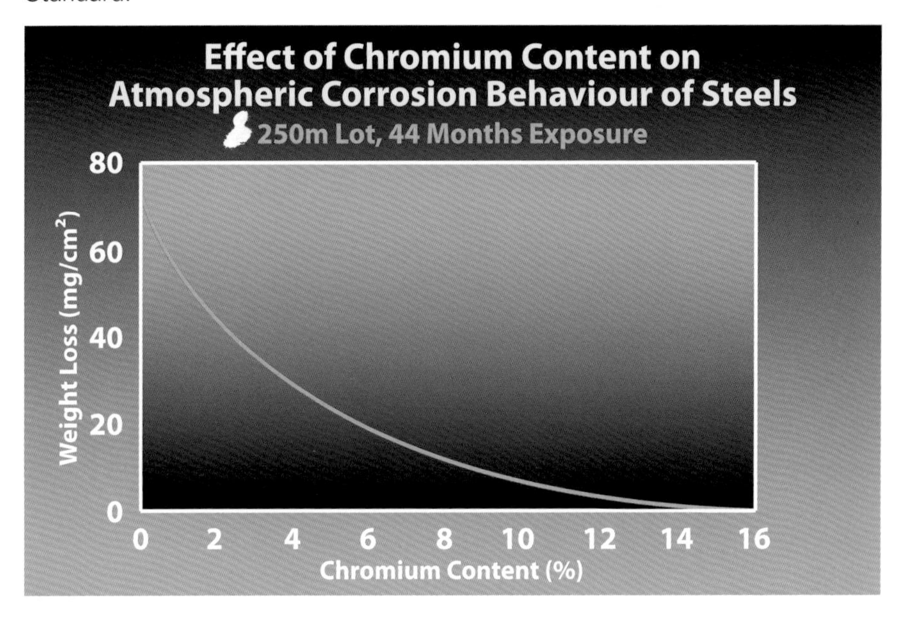

So why does chromium give this improvement. It's all down to the "passive film" that forms on steels with enough chromium. When ordinary steel corrodes it forms rust or iron oxide (Fe_2O_3) with the oxygen from the air. This is porous and flakes off, allowing further corrosion to take place.

Typical appearance of rusted carbon steel

In stainless steels, the oxygen from the air forms a very thin layer of chromium oxide (Cr_2O_3). This layer of oxide is not porous and prevents further oxidation of the steel surface. If it is damaged it immediately self-repairs.

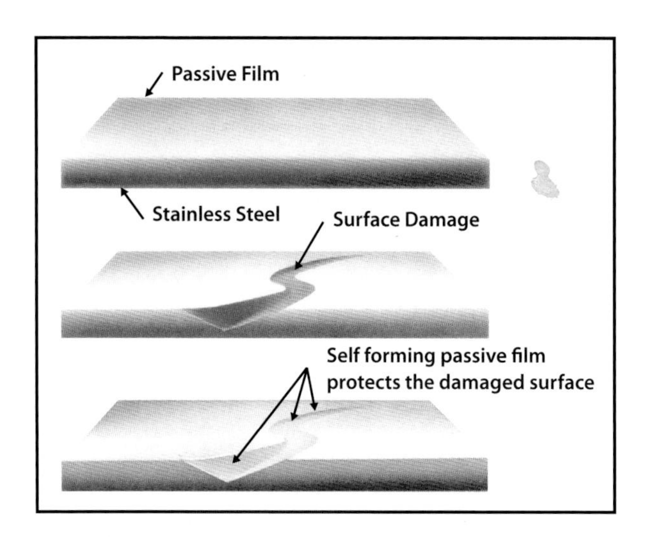

Mechanism of self-repairing passive film on stainless steel

This passive film is only a few nanometres thick. A nanometre is a millionth of a millimetre. This means that you can see through the oxide film to the metallic surface underneath.

Another way of looking at this to say that the passive film is one ten-thousandth the thickness of a human hair.

The passive film can be attacked as we shall see in chapter 10 but for most normal environments it is sufficient to prevent corrosion.

Chapter 5 - Why Use a Stainless Steel

Obviously, stainless steel is mostly used for its corrosion resistance. However, there are other properties which can be equally important. These include:

- **Attractive appearance, wide range of surface finishes**

There are mill finishes, bright polished, dull polished, patterned, coloured and a virtually unlimited combination of these types to give designers a wide choice of final appearance.

- **Ease of cleaning, hygienic - no bug traps**

It has been demonstrated in scientific tests that stainless steel can be cleaned efficiently even after prolonged use and normal wear and tear. This has led to it being a material of choice for the catering and hospital sectors.

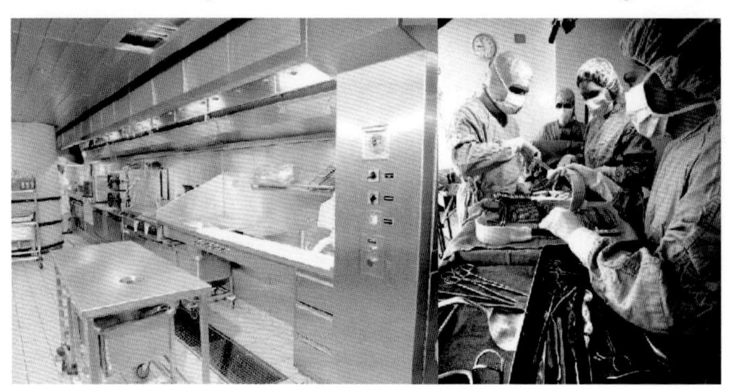

- **Surface coatings not required**

Unlike surface coatings, stainless steel is 100% stainless "all the way through". This is in contrast to materials such as galvanised steel where the underlying steel is exposed once the coating is damaged.

- **Formability**

In many applications it is necessary for complex shapes to be formed. With the appropriate addition of elements such as nickel, the formability of stainless steels allows the manufacture of such articles as sinks and saucepans.

- **Weldability**

For many applications, it is necessary to be able to weld stainless steel together. In thin sections, most stainless steels are weldable. However, for thick sections it is necessary to add elements such as nickel and to use the low carbon "L" variant to improve weldability to make products such as pressure vessels and process piping.

- **Magnetic properties**

For most applications, the non-magnetic property of austenitic stainless steels is an incidental one. However, for a small number of highly technical applications the non-magnetic feature of stainless steel is the dominant one. MRI Body Scanners are typical of this type.

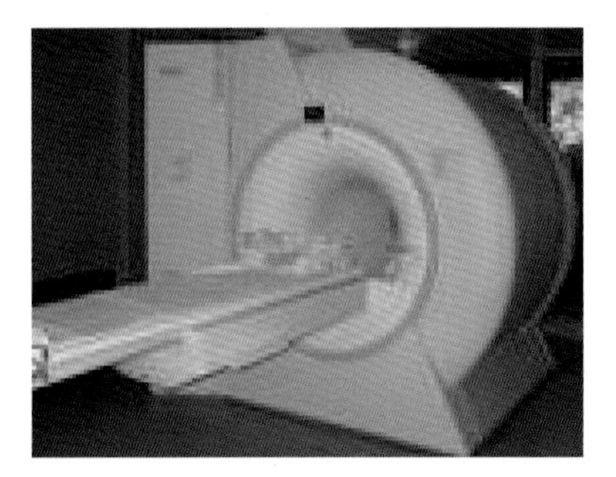

Some applications for stainless steel require that they be magnetic, for example induction-heated saucepans, magnetically activated valves.

- **High temperature properties**

The passive oxide film allows stainless steels to withstand the effects of hot gases. This explains the use of stainless steel in furnace applications. Some stainless steels can be specified for up to 1150° C.

- **Low temperature properties**

Most steels cannot be used at low temperatures (below about -50° C) as they become brittle. However, some austenitic stainless steels can be used as low as -270° C, for example for storing liquefied gases such as helium.

The family of stainless steels can therefore be used over a wide range of temperatures and conditions. The unique combination of properties explains why stainless steel is growing at such a high rate.

Chapter 6 - The Structure Stainless Steel

This chapter is about as technical as it gets, so if you get through this everything else will be plain sailing!

You will recall that ordinary steel is an alloy of iron and carbon. At normal temperatures, iron atoms are arranged in a pattern or lattice as shown below:

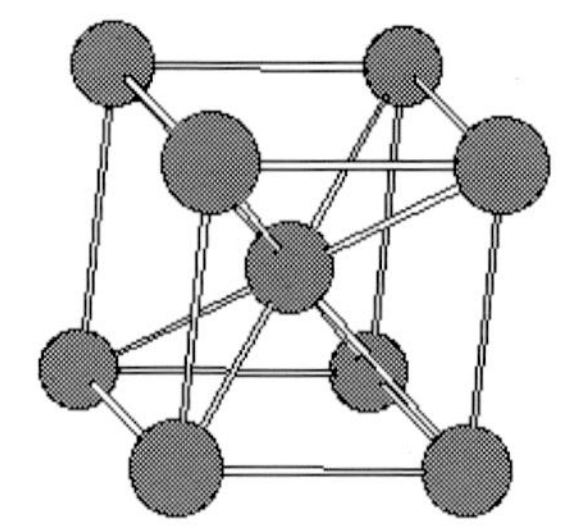

This is called a Body Centred Cubic (BCC) structure for fairly obvious reasons. There is an atom at each corner of a cube and one in the middle. The arrangement of atoms is called a lattice. The atoms of carbon are smaller and have to fit in as best they can within the structure. In iron and steel this structure is also called ferritic.

The ferritic structure is magnetic.

When ordinary steels are heated up to about 900° C, the atomic structure changes to this pattern:

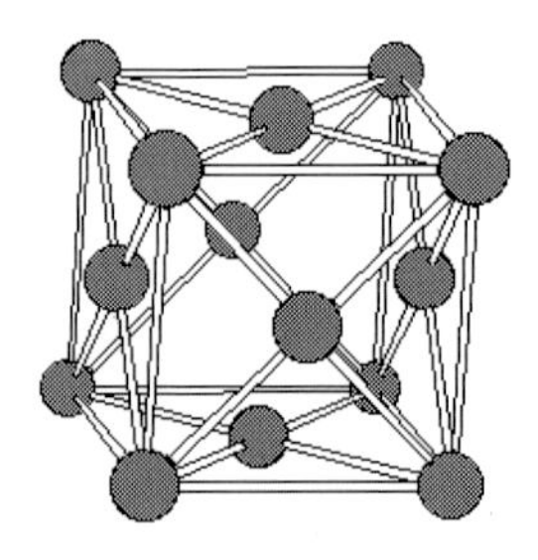

This is called Face Centred Cubic (FCC). There is one atom at each corner and one in the middle of each face. This structure is also called austenitic.

The austenitic structure is non-magnetic.

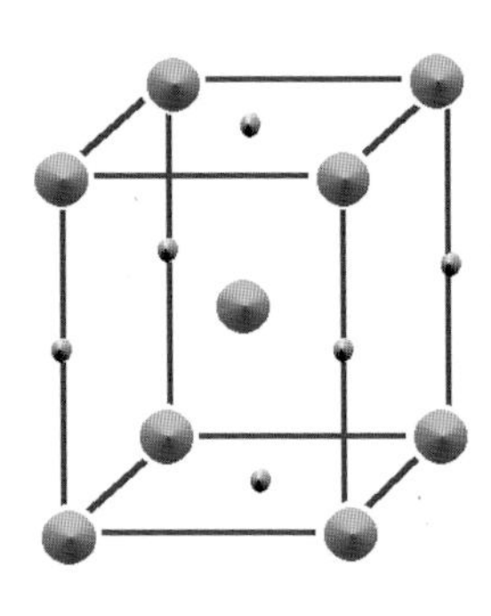

In an ordinary steel, when you cool it down again, the structure changes back to ferritic OR if you cool it fast enough by, for example, quenching it in water or oil it changes to another pattern:

This is called Body Centred Tetragonal or martensitic. It's a bit like the BCC structure but slightly distorted by the small carbon atoms in one direction. Martensite in the as-quenched condition is hard but brittle. To make it useful it is tempered at up to about 700° C. This gives the steel a good combination of strength and toughness.

So what has this stuff about ordinary steels got to do with stainless steels?

Armed only with this basic knowledge, we can now begin to understand the different types of stainless steel.

Ferritic Stainless Steels

Stainless steels which only have chromium additions like type 409 (12% Cr) and 430 (17% Cr) retain the ferritic structure. Chromium and iron atoms are similar in size, so they fit quite well into the iron lattice. Chromium itself also has a BCC ferritic structure.

These steels are magnetic like ordinary steels because they have the ferritic structure.

Austenitic Stainless Steels

In carbon and low alloy steels, the austenitic structure only exists at high temperature above about 900 °C. In stainless steels, if you add elements which favour the FCC austenitic structure like nickel, manganese and copper the steel stays austenitic all the way down to normal temperatures. So, with 8% nickel, 1.4301 (304) is an austenitic steel.

These stainless steels are non-magnetic because austenite is non-magnetic.

Martensitic Stainless Steels

These work like carbon and low-alloy steels as they can be quenched and tempered except that the chromium content makes them corrosion resistant as well.

These stainless steels are magnetic because they have the martensitic structure.

Duplex Stainless Steels (Ferritic-Austenitic)

These stainless steels are deliberately balanced in composition to give approximately 50% austenite and 50% ferrite.

The ferrite content makes them magnetic but not so much as the ferritic or martensitic steels due to the 50% austenite.

The duplex structure makes these steels stronger than either ferritic or austenitic steels.

Precipitation Hardening Steels

These are a specialist type which work by using elements like copper (Cu) and aluminium (Al) to produce tiny particles on heat treatment. These particles act as a strengthening agent in the steel.

So to summarise this chapter:

- Carbon and low-alloy steels are ferritic at room temperature BUT
- Transform to austenite at high temperature AND
- Transform back to ferrite when cooled OR
- If cooled fast enough, transform to martensite

TEMP ↑	SLOW COOL	FAST COOL
	AUSTENITE	AUSTENITE
	FERRITE & AUSTENITE	
	FERRITE	MARTENSITE

- Stainless steels which contain only chromium as the main alloying element are either ferritic or martensitic BUT
- If you add enough of elements such as nickel, manganese or copper THEN
- You can have austenitic structure at room temperature
- Also possible for mixed austenitic and ferritic steels called duplex by balancing the amounts of austenite and ferrite forming elements

Now we can begin to see why the different types are used for which application.

Chapter 7 - Types of Stainless Steel

The five basic types described in chapter 6 are used in approximately the following proportions:

- Austenitic 65 - 70%
- Ferritic 20 - 25%%
- Martensitic about 7%
- Duplex about 1%
- Precipitation Hardening about 2%

PROPERTIES OF AUSTENITIC TYPE

Pros	Cons	Typical Applications
Easy to produce Formable – stretch forming Weldable in thick sections Low temperature toughness Oxidation resistance Non-magnetic Strengthened by cold work Can be surface hardened High alloy grades giving high level of corrosion resistance	Subject to big price swings due to nickel volatility High alloy grades more expensive Not heat treatable in bulk Low thermal conductivity High thermal expansion Difficult to machine	Sinks, saucepans, cutlery, cladding, handrails, roofing, catering surfaces, chemical, pharmaceutical, pressure vessels, food processing, oil and gas, street furniture, sanitary equipment, hospital equipment, MRI scanners, building products e.g. wall ties, furnaces, electrical energy, cryogenic storage vessels, springs, rail carriages, high spec exhaust systems, automotive structural (under development), process piping, medical devices, water tubing, nuclear processing, yacht trim

PROPERTIES OF FERRITIC TYPE

Pros	Cons	Typical Applications
Formable – deep drawing Oxidation resistance High alloy grades giving high level of corrosion resistance Price stability – very low or no nickel Similar thermal properties to carbon steels	Not weldable in thick sections Not as stretch formable as austenitic Not heat-treatable Poor low temperature toughness	Washing machine drums, automotive exhaust systems, catering, microwave oven linings, cutlery, hot water tanks, internal decorative tubing, automotive trim, induction heating saucepans, window hinges, ventilation ducting, lift panels, electrical enclosures, coal wagons, food handling e.g. sugar beet, containers, bus chassis, furnaces, solenoids

PROPERTIES OF MARTENSITIC TYPE

Pros	Cons	Typical Applications
High strength with moderate toughness at room temperature Good for blades Price stability – very low or no nickel Similar thermal properties to carbon steels	Poor weldability (except for low C grades) Poor low temperature toughness Process route far more complex than austenitic Limited corrosion resistance	Razor strip, high quality knife blades, scalpels, shafts, hydraulic rams, wear resistant plate, materials handling, valves, bearings, oil and gas, glass furnace rolls, high temperature bolting, fibre board plates, springs, fasteners, garden tools

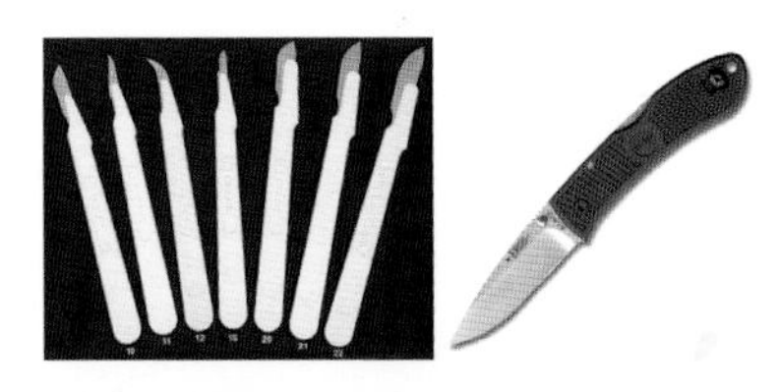

PROPERTIES OF DUPLEX TYPE

Pros	Cons	Typical Applications
Improved strength compared to austenitic Moderate low temperature toughness Weldable in thick sections High alloy grades giving high level of corrosion resistance Better price stability than austenitic Can be cold worked	Not as easy to make, phase balance More care in welding Not easily formed Difficult to machine	Chemical processing, sub-sea oil and gas, structural applications, bridges, hot water tanks, pulp and paper, desalination plants, water treatment, processing aggressive foods, brewing vessels, fasteners, chemical transport

PROPERTIES OF PRECIPITATION HARDENING TYPE

Pros	Cons	Typical Applications
Very high strength and better toughness than martensitic Better corrosion resistance than martensitic	Quite expensive Complex process route Not easily weldable Not formable	Pump shafts, valves, aerospace, springs

We shall now have a look at some of the individual grades of steel which belong to each of the basic types.

Chapter 8 - Grades of Stainless Steel

Although there are only 5 basic types of stainless steel, there are almost 200 grades listed in the main European standards. Of course, not all of these grades are commonly used but in theory a grade can only be included if it is sold in reasonable amounts. There have been various attempts to define a logical coding system for stainless steels. The most well known one is the AISI system. This system uses three series:

200 – Austenitic chromium manganese nickel stainless steels
300 – Austenitic chromium nickel stainless steels
400 – Ferritic and martensitic chromium stainless steels

Familiar grades from this system include 201, 202, 304/304L, 316/316L, 430, 420. Duplex and precipitation hardening steels have not fitted into the system.

The modern US system is the UNS (Unified Numbering System). This uses the AISI system as a starting point and then adds 2 more digits to further define the steel. For example, S31600 is 316 and S31603. The initial S means stainless.

The AISI system was the basis for the British Standard system. Typical examples were 304S11, 304S31, 316S11, 316S31, 431S29. The "S" meant "stainless" to distinguish them from other types of steel. The last two digits were effectively arbitrary. However, 11 usually meant "L" grade and 31 meant normal.

The modern European system is based on the German DIN system and was agreed by a committee representing all the European countries. Although the EN standards for flat and long products (bar rod and sections) have been around since 1995, the grade names and numbers are still relatively unfamiliar. The benefit is that there is a certain logic in the system which is why it was chosen. As an example, consider the most familiar stainless steel – type 304. In EN terms this is defined as:

EN Name – X5CrNi18-10 EN Number – 1.4301

In the Name, the initial "X" always means "stainless or heat resisting steel". The next digits refer to the approximate carbon content. So "5" means 0.05%, 12 means 0.12%, 105 means 1.05%. Then come the important elements and their approximate contents. So X5CrNi18-10 has about 18% Cr and 10% Ni.

There is also some logic in the EN Number:

1.40xx and 1.41xx – ferritic and martensitic stainless steels
1.43xx – stainless steel without Mo (both austenitic and duplex)
1.44xx – stainless steel with Mo (both austenitic and duplex)
1.45xx – stainless steels with special additions (ferritic, austenitic and duplex)
1.47xx – ferritic heat resisting steels
1.48xx – austenitic heat resisting steels
1.49xx – creep resisting steels

The following tables show some typical grades of each of the 5 main types.

SOME EXAMPLES OF AUSTENITIC GRADES

EN name	EN number	Chemical Composition from EN 10088/EN 10095 (single values are maximum)				
		C	Cr	Mo	NI	Others
X12CrMnNi17-7-5	1.4372	0.15	16.0/18.0		3.5/5.5	Mn 5.5/7.5
X10CrNi18-8	1.4310	0.05/0.15	16.0/19.0		6.0/9.5	
X5CrNi18-10	1.4301	0.07	17.5/19.5		8.0/10.5	
X2CrNi18-9	1.4307	0.030	17.5/19.5		8.0/10.5	
X8CrNiS18-9	1.4305	0.10	17.0/19.0		8.0/10.0	S: 0.15/0.35 Cu:1.00
X5CrNiMo17-12-2	1.4401	0.07	16.5/18.5	2.00/2.50	10.0/13.0	
X2CrNiMo17-12-2	1.4404	0.030	16.5/18.5	2.00/2.50	10.0/13.0	
X8CrNi25-21	1.4845	0.10	24.0/26.0		19.0/22.0	
X1NiCrMoCu25-20-5	1.4539	0.020	19.0/21.0	4.0/5.0	24.0/26.0	
X1NiCrMoCuN20-18-7	1.4547	0.020	19.5/20.5	6.0/7.0	17.5/18.5	N 0.18/0.25

EN number	'Old' BS no.	UNS number	Common name	Applications
1.4372	-	S20100	201	Lighting columns
1.4310	301S21	S30100	301	Springs
1.4301	304S31	S30400	304	Sinks
1.4307	304S11	S30403	304L	Pressure vessels
1.4305	303S31	S30300	303	Free machined components
1.4401	316S31	S31600	316	Architectural cladding
1.4404	316S11	S31603	316L	Pharmaceuticals, oil and gas
1.4845	310S24	S31000	310	High temperature furnace
1.4539	904S13	N08904	904L	Sulphuric acid service
1.4547	-	S31254	254SMO	Severely corrosive environments, desalination, pulp and paper

SOME EXAMPLES OF FERRITIC GRADES

EN name	EN number	Chemical Composition from EN 10088 (single values are maximum)				
		C	Cr	Mo	NI	Others
X2CrNi12	1.4003	0.030	10.5/12.5		0.30/1.00	
X6Cr13	1.4000	0.08	12.0/14.0			
X2CrTi12	1.4512	0.030	10.5/12.5			Ti: 6(C+N)/0.65
X6Cr17	1.4016	0.08	16.0/18.0			
X3CrTi17	1.4510	0.05	16.0/18.0			Ti: 4(C+N)+0.15/0.80
X2CrTiNb18	1.4509	0.030	17.5/18.5			Ti: 0.10/0.60 Nb: 3C+0.30/1.00
X6CrMoTi18-2	1.4521	0.025	17.0/20.0	1.80/2.50		Ti: 4(C+N)+0.15/0.80
X2CrMoTi29-4	1.4592	0.025	28.0/30.0	3.50/4.50		Ti: 4(C+N)+0.15/0.80

EN number	'Old' BS no.	UNS number	Common name	Applications
1.4003	-	S40977	3CR12	Coal wagons, bus chassis
1.4000	-	S41008	410S	Petroleum refining
1.4512	409S19	S40900	409	Exhaust systems
1.4016	430S17	S43000	430	Washing machine drums
1.4510	-	S43035	430Ti(439)	Hot water tanks, exhaust systems
1.4509	-	S43932	441	Condensation boilers
1.4521	-	S44400	444	Heat exchanger tubing
1.4592	-	S44700	29-4	Seawater applications

SOME EXAMPLES OF DUPLEX GRADES

EN name	EN number	Chemical Composition from EN 10088 (single values are maximum)				
		C	Cr	Mo	NI	Others
X2CrMnNiMoN21-5-1	1.4162					
X5CrNiN23-4	1.4362	0.030	22.0/24.0	0.10/0.60	3.5/5.5	Cu: 0.10/0.60 N: 0.05/0.20
X2CrNiMoN22-5-3	1.4462	0.030	21.0/23.0	2.50/3.50	4.5/6.5	N: 0.10/0.22
X5CrNiMoN25-7-4	1.4410	0.030	24.0/26.0	3.0/4.5	6.0/8.0	
X2CrNiMoCuWN25-7-4	1.4501	0.030	24.0/26.0	3.0/4.0	6.0/8.0	Cu: 0.50/1.00 W: 0.50/1.00

EN number	'Old' BS no.	UNS number	Common name	Applications
1.4162	-	S32101	2101LDX	Rebar
1.4362	-	S32304	2304	Domestic hot water tanks
-	-	S32003	AL 2003	Water treatment
1.4462	318S13	S32205	2205	Chemical processing
1.4410	-	S32750	2507	Subsea oil and gas
1.4501	-	S32760	Zeron 100	Subsea oil and gas
-	-	S32707	2707 HD	Seawater cooled heat exchanger

SOME EXAMPLES OF MARTENSITIC GRADES

EN name	EN number	Chemical Composition from EN 10088/EN 10302 (single values are maximum)				
		C	Cr	Mo	NI	Others
X12CrS13	1.4005	0.06/0.15	12.0/14.0			S: 0.15/0.35
X12Cr13	1.4006	0.08/0.15	11.5/13.5			
X20Cr13	1.4021	0.16/0.25	12.0/14.0			
X17CrNi16-2	1.4057	0.12/0.22	15.0/17.0		1.50/2.50	
X3CrNiMo13-4	1.4313	0.05	12.0/14.0	0.30/0.70	3.5/4.5	
X105CrMo17	1.4125	0.95/1.20	16.0/18.0	0.40/0.80		
X12CrNiMoV12-3	1.4938	0.08/0.15	11.0/12.5	1.50/2.00	2.00/3.00	N: 0.020/0.040 V: 0.25/0.40

EN number	'Old' BS no.	UNS number	Common name	Applications
1.4005	416S21	S41600	416	Gears, valves, high volume machined components
1.4006	410S21	S41000	410	Chemical engineering
1.4021	420S29	S42000	420	Surgical instruments, kitchen knives
1.4057	431S37	S43100	431	Propellor shaft
1.4313	-	S41500	F6NM	Offshore oil and gas
1.4125	-	S44004	440C	High wear resistant components
1.4938	-	S64152	Jethete M152	Power generation

SOME EXAMPLES OF PRECIPITATION HARDENING GRADES

EN name	EN number	Chemical Composition from EN 10088 (single values are maximum)				
		C	Cr	Mo	NI	Others
X5CrNiCuNb16-4	1.4542	0.07	15.0/17.0		3.0/5.0	Cu: 3.0/5.0 Nb: 5xC/0.45
X7CrNiAl17-7	1.4568	0.09	16.0/18.0		6.5/7.8	Al: 0.70/1.50
X5CrNiMoCuNb14-5	1.4594	0.07	13.0/15.0	1.20/2.00	5.0/6.0	Cu: 1.20/2.00 Nb: 0.15/0.60

EN number	'Old' BS no.	UNS number	Common name	Applications
1.4542	-	S17400	17-4 PH	Turbine blades, plastic moulding dies
1.4568	-	S17700	17-7 PH	High strength springs
1.4594	S143/4/5	-	FV 520B	Aerospace

Chapter 9 - Magnetic Properties of Stainless Steel

Now that we have covered the basic structure and properties of stainless steels, it is worth spending a little time on this subject. This is particularly important as there is a lot of mis-information around.

"A magnet will not stick to 304 stainless steel but it will to type 430 or to any other inferior material".

Although the statement is generally correct, the truth is rather more complex.

Recap from Chapter 6:

Steel Structure	Magnetic Properties
Ferritic (Body Centred Cubic)	Magnetic
Martensitic (Body Centred Tetragonal)	Magnetic
Austenitic (Face Centred Cubic)	Non-Magnetic
Duplex (Mixed Austenitic/Ferritic)	Magnetic

Austenitic stainless steels such as type 1.4301 (304), 1.4401 (316) are nominally non-magnetic because the austenite structure is non-magnetic. However, there are two reasons why an austenitic stainless steel can have some degree of magnetic response.

Effect of Ferrite

All austenitic stainless steels contain a small amount of ferrite. Usually, this is not enough to attract a normal magnet. However, if the balance of elements in the steel favours the ferritic end of the spectrum, it is possible for the amount of ferrite to be sufficient to cause a significant magnetic response. Also, some types of product are deliberately balanced to have a significant amount of ferrite. Castings are in this category and normally have about 10% ferrite.
Welding can also induce a greater magnetic response in the melted zone where ferrite is produced in greater quantities than in the parent material.

Effect of Martensite

Another reason for a magnetic response is the transformation of some of the austenite to the magnetic martensite phase. This can occur by either cold working or cooling the steel to sub-zero temperatures. Austenitic steels with the lowest alloy contents are more susceptible to this kind of transformation. This is why steels like type 301 with only 6.0% minimum nickel can be quite magnetic after cold working. Some steels have been developed specifically to remain non-magnetic even after a high degree of cold working and/or cooling to very low temperatures. These include steels like 304LN and 316LN. Type 310 is a high temperature steel but its high alloy content also makes it resistant to martensite transformation and is therefore also used as a non-magnetic steel.

Some products are more susceptible to this effect of cold work than others. For example, cold drawn bar is likely to be more magnetic than cold rolled 2B sheet as the final operation puts a fair amount of work into the bar.

You can verify the effect of cold working at home with a fridge magnet and your stainless steel sink. The magnet will not stick to the flat drainer. However, as you move the magnet into the bowl especially into the corners, you will find that there is quite a strong attraction.

Magnetic Permeability

The magnetic properties of stainless steel are often expressed by using the relative magnetic permeability. A completely non-magnetic material has a relative magnetic permeability of 1. Standard austenitic grades in the fully softened condition have a typical magnetic permeability of 1.02-1.1. In some cases, the amount of ferrite may be enough to attract a strong magnet. With cold working, the low alloy grades can quickly be increased to values as high as 6. The most non-magnetic grades are designed to have a maximum permeability as low as 1.005 whatever processing they undergo. For comparison, ferritic grades have a magnetic permeability of around 200.

Chapter 10 - The Testing of Stainless Steels

In this chapter, we will take a brief look at the tests that are carried out on stainless steels before being released into the supply chain. The information will help you interpret the test certificates which are often supplied with the material.

Unfortunately, there is no agreed format or content for test certificates so it is sometimes a little difficult to find all the information you might want.

Before looking at an actual test certificate, we need to understand what kind of information is being presented.

All steel products are tested to ensure conformance with recognised standards. A typical test certificate always shows:

- Standards and grades
- Product description (dimensions, finish etc)
- Chemical composition
- Room temperature tensile test properties
- Hardness

It may show:
- Tolerance standard
- Impact toughness
- High temperature tensile test properties
- Corrosion tests
- NDT (Non Destructive Tests) e.g. ultrasonic, dye penetrant, magnetic particle inspection, eddy current testing

Standards and Grades
The table shows some common examples of standards and grades.

Standard	Products Covered	Common Grades
EN 10088-2	Flat Products	1.4301 1.4307 1.4401 1.4404
EN 10088-3	Long Products	1.4301 1.4307 1.4401 1.4404
EN 10095	Heat Resisting Steels	1.4845 1.4835
ASTM A240	Flat Products	304 304L 316 316L
ASTM A276	Long Products	304 304L 316 316L

There can also be company specific standards, for example Rolls Royce. These usually call up a basic standard like EN or ASTM and then add more requirements. EN (European Norm) is the designation for standards issued by the European Standards body CEN. Each national standards body issues its own language version of the EN standard. BS EN standards are issued in English, DIN EN is German etc. The technical content of the standards is identical. ASTM is the American Society for the Testing of Metals.

Chemical Composition

There is no standard for which elements must be shown on a test certificate. The elements which are specified in the material standard must be shown but each producer has its own rules for which other elements are shown.

For example, grade 1.4301 does not have any deliberate addition of Mo. Some mills choose not to show the Mo content for this grade, whereas others will have a fixed number of elements which are always shown.

Some producers show the composition range required by the standard but some choose to show only the actual composition.

Room Temperature Tensile Properties

All stainless steel is tested in this way. The main purpose is to show that the heat treatment has been carried out in the correct manner. This ensures that the steel will be in the correct condition for further processing. It is particularly important when the end product is made by forming. Steel which has not been softened properly may fail when bending, drawing or stretch forming is being done.

The frequency of testing is specified in the standards. A typical frequency is at each end of a coil or plate.

A tensile test piece looks like this for a strip or sheet product

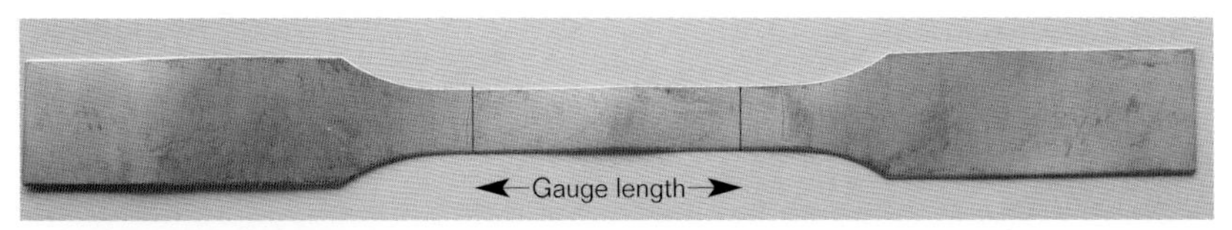

For bar or thick plate, a round test piece would be machined but still having the same shape with the centre being narrower than the ends. The elongation is measured on a standard length called the gauge length typically 50mm marked on the parallel portion of the test piece.

The test piece is then loaded into a tensile test machine and put into the jaws

The piece is then slowly stretched until it breaks. Three properties are measured in a tensile test:

- 0.2% Proof stress
- UTS (Ultimate tensile stress)
- Elongation

On a test certificate these values are usually designated as:

$R_{p0.2}$ = 0.2% Proof Stress
R_m = Ultimate Tensile Stress
A = Elongation

Before going any further, we need to define some basic concepts:

Stress – This is a measure of how much load is being applied to the test piece per unit area. The most common measures, at least on European test certificates, are N/mm^2 (Newtons per square millimetre) and MPa (Megapascals). These are identical. To give you some sort of picture of what this means, imagine a small apple hanging on a thin wire measuring 1mm square. This is about 1 N/mm^2 or 1 MPa. Another way of picturing this is to say that a bag of sugar hanging on a 1mm square wire is about 10 N/mm^2 or 10 MPa.

Strain – This is a measure of how much the test piece is being stretched by the load. So 0.2% strain means that the gauge length of 50 mm has been stretched to 50.1 mm.

As the test piece is stretched the machine automatically produces a stress-strain curve. This curve is used to calculate the three basic measures. The diagram below shows a typical curve:

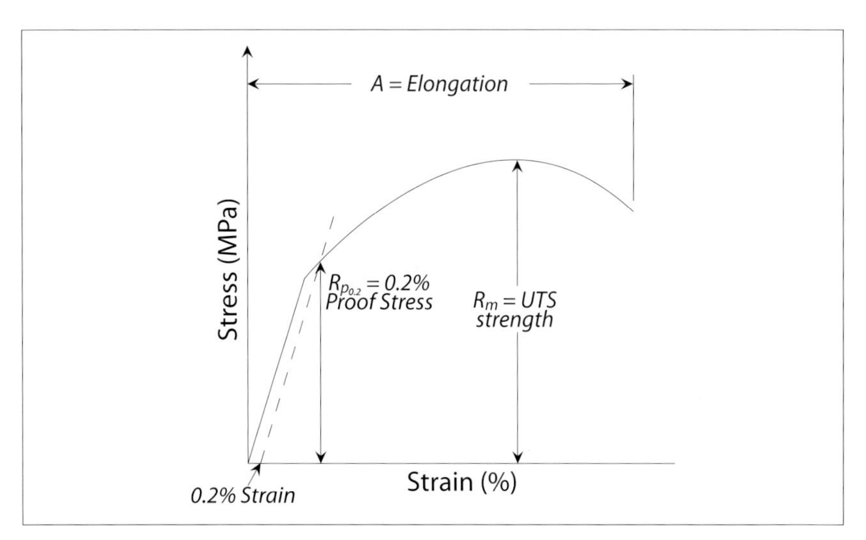

Typical values for a standard austenitic stainless steel are:

$R_{p0.2}$ = 300 MPa
R_m = 580 MPa
A = 50%

So returning to the "apple" picture, it takes about 300 apples hanging on the 1mm square wire to produce a strain of 0.2%. Then another 280 apples to get to the UTS. The high elongation of 50% is an indication of good ductility.

Each type of steel has its own characteristic stress/strain curve:

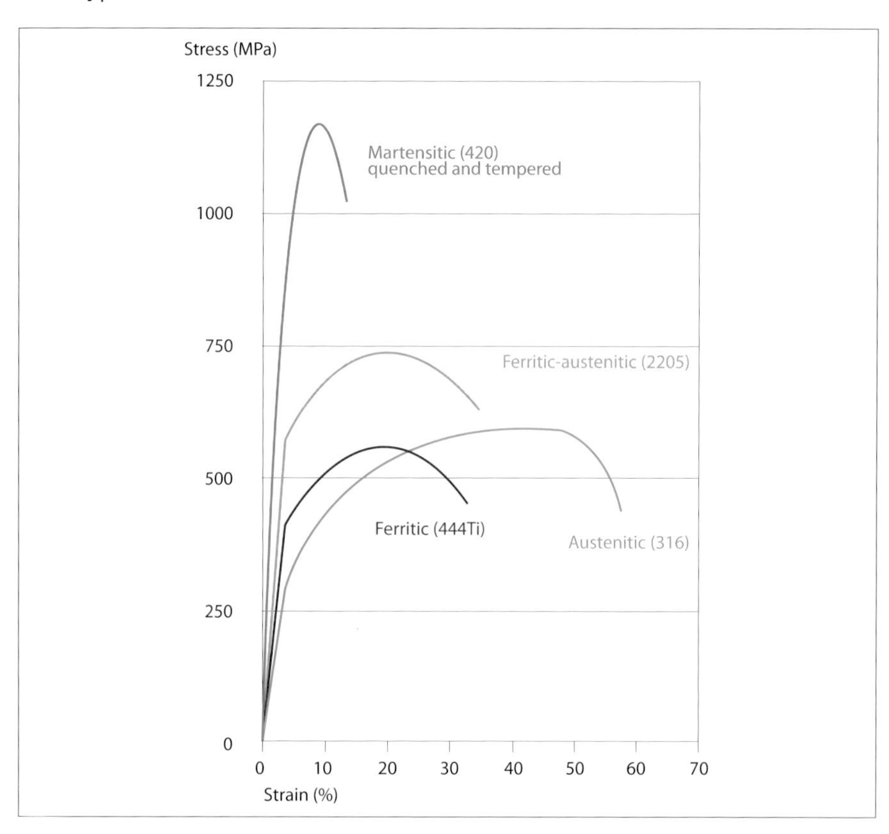

Austenitic steels have the best elongation of all stainless steel types. This gives them their excellent stretch forming properties.

Martensitic steels are very strong but have a low elongation. This explains why martensitic steels cannot be formed easily.

Ferritic-austenitic (Duplex) steels are stronger than standard austenitic steels but have lower elongation values. They are therefore moderately formable compared to austenitic steels.

Ferritic steels have a higher proof strength than austenitic steels but a lower UTS as they do not work harden so much. Elongation values are lower so they are less formable in stretch forming than austenitics. However, they can be deep drawn very well as this operation does not involve stretching.

These curves show the steels in their "normal" condition. This means:

Martensitic – Quench and tempered – The tempering process allows a wide range of strengths to be achieved in the same steel. The graph below shows how the strength varies with the tempering temperature:

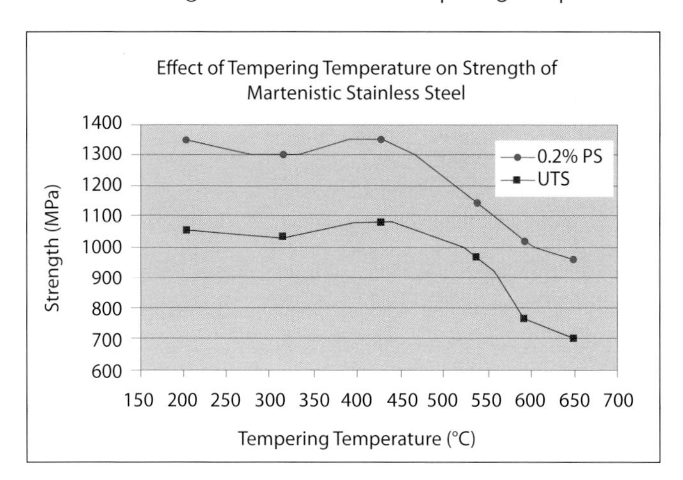

Ferritic, austenitic, duplex – Fully softened – Ferritic, austenitic and duplex steels can be work hardened by cold rolling as strip or cold drawing as bar to enhance their strength. Austenitics are particularly used in this way as they can be work hardened and still retain adequate ductility. Ferritic steels are not much used in the work hardened condition as they have inadequate ductility. Duplex steels can be usefully work hardened and as they have higher starting strength than austenitic steels, some very high values can be obtained.

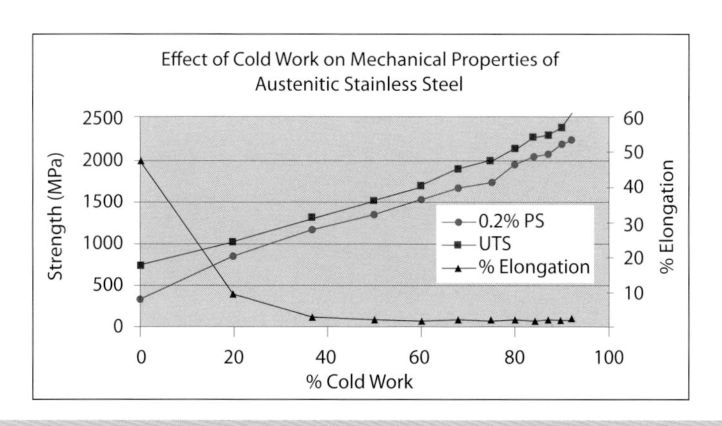

The following table is a rough guide to what strength can be obtained for each type of steel.

Steel Type	Strengthening Method	Maximum strength (MPa)
Duplex	Microstructure	700
Duplex	Work Hardening	3000 (fine wire)
Austenitic	Work hardening	1800
PH	Aging	2400
Martensitic	Quench and tempering	1800
Austenitic and Duplex	Surface hardening	1200 HV

The maximum strength achievable is very much dependent on the thickness of diameter of the product in question. The thinner the material, the higher the strength can be.

The last method of surface hardening is used where it is only necessary for a thin layer of the surface to have a high strength. In this case, you can't measure strength directly with a tensile test piece as the layer may only be 20 micron (0.02 mm) thick.

In this case, hardness testing is used to measure the thin layer.

Hardness Testing

It is usual for a hardness check to be carried out on the tensile test piece before it is tested. Most standards have maximum hardness values to ensure that the steel is sufficiently soft for further processing.

Essentially, the test measures the ability of a material to resist indentation by a hard piece of material. This varies with test method:

Brinell (HB) – This test uses a hardened steel ball
Vickers (HV) – This uses a pyramid shaped diamond
Rockwell A (HRA) – A cone shaped diamond
Rockwell B (HRB) – A hardened steel ball or diamond ball
Rockwell C (HRC) – A cone shaped diamond

Brinell and Vickers tests tend to be used in Europe, whereas Rockwell is more American. Although the details of the methods vary, the basic idea is the same. The indenter is pressed into the surface of the material under a specified load and the cross section of the indentation is measured. The wider the indentation, the softer is the material.

The load is applied to the pyramidal indenter and then the diagonals of the impression are measured.

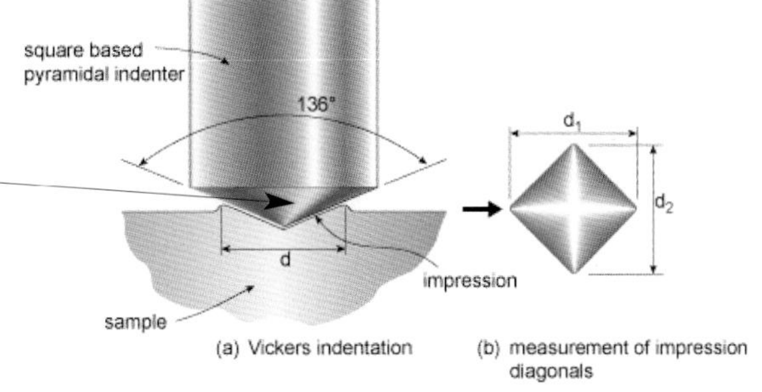

(a) Vickers indentation (b) measurement of impression diagonals

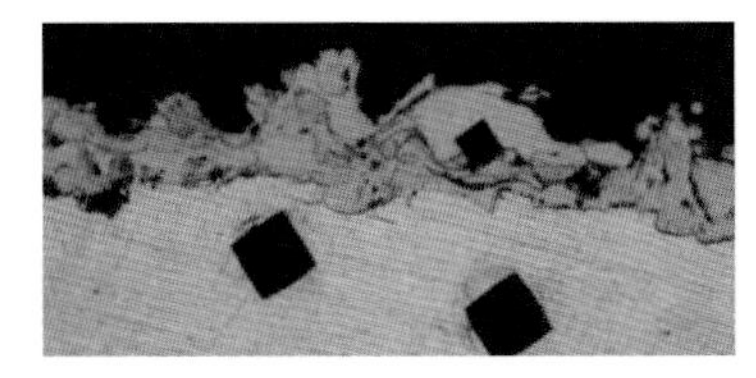

Real hardness indentations under the microscope. Measurements are taken across the diagonals and the hardness value is converted from a table.

The hardness is related to the UTS and there are conversion tables which show equivalent values of other hardness methods and to UTS. These tables can only be used for guidance.

Impact Toughness

In everyday conversation, the words "strong" and "tough" are used more or less interchangeably. In the testing of materials, they mean something quite different.

Whereas tensile strength is the ability to withstand a "slow pull", toughness is the ability to withstand a "sharp blow". The two do not necessarily go together. Think of a glass bottle. Imagine trying to pull it apart with your bare hands. Glass is actually quite strong. But it is quite easy to break glass if you just drop it on the floor. Glass is brittle or has low "impact toughness".

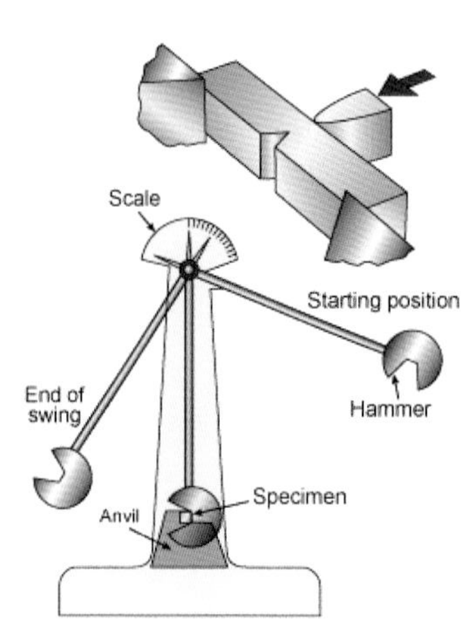

On the other hand, steels can be both strong and tough. Toughness is measured by taking a small sample and machining a notch in it. It is then placed in the path of a hammer. The hammer swings into the sample and breaks it. The position where the hammer swings up the other side is noted. The further the hammer swings through, the lower is the toughness. The impact toughness is measured in Joules or ft-lb (foot-pounds). This is a measure of how much energy is absorbed by the test piece. Sometimes the test piece does not break and the hammer is completely stopped by the sample. This is technically an invalid test but shows very high toughness.

Impact testing is an optional test and is only carried out where the material standard requires it or where the application demands it.

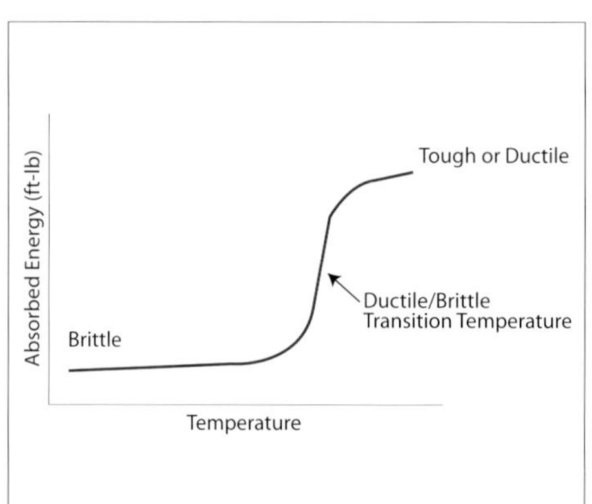

One of the most important aspects of steels generally is that they lose toughness at low temperatures. This loss of toughness is quite sudden as shown by the following schematic diagram.

The temperature at which the sudden loss occurs is called the Ductile/Brittle Transition Temperature (DBTT). The actual temperature varies with the grade of steel.

At the top of the curve, the steel is 100% tough/ductile. At the bottom it is 100% brittle. Somewhere on the steep part of the curve, the fracture is 50% of each type. The temperature where this occurs is the Ductile/Brittle Transition Temperature (DBTT). It is measured by carrying out impact tests at a range of temperature and plotting the impact behaviour.

The ductile to brittle transition occurs in all steels with a martensitic or ferritic structure including stainless steels. It also applies to duplex steels which has a significant ferritic component.

However, austenitic stainless steels do not show this kind of behaviour. The impact toughness falls off gradually with temperature and the failure mode remains ductile.

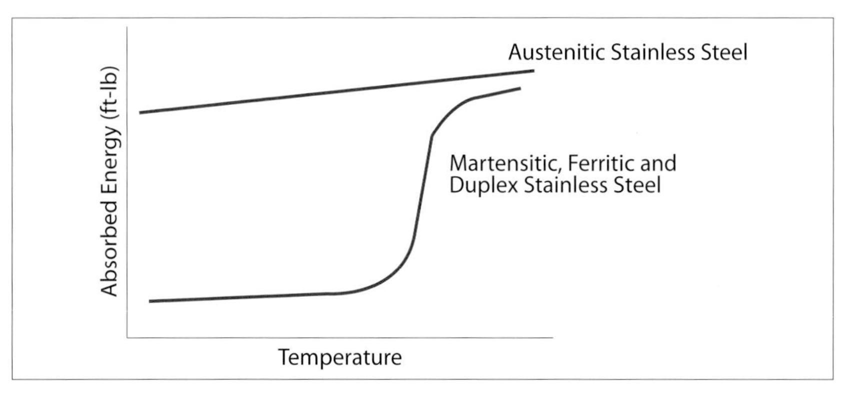

Impact testing is most often required in certain sectors like:

- Oil and gas
- Pressure vessels
- Low temperature
- Structural

To avoid the many tests required for a full transition curve, it is common practice to carry out tests at a single low temperature. For example, many specifications for offshore oil and gas require an impact tests at minus 50 deg C. This gives a wide margin of safety since it doesn't get quite as cold as this in the North Sea.

CERTIFICATE	CERT NO	INTERNATIONAL STAINLESS STEEL COMPANY
Date	X28238240	
12/01/2009		

CUSTOMER	CUSTOMER ORDER	STANDARDS AND GRADES
ACME STAINLESS STEEL 55 CLARKEHOUSE RD SHEFFIELD S10 2LE	X6715670	EN 10028-7 1.4462 EN 10088-2 1.4462 ASTM A240 S31803 ASME SA240 S31803 TOLERANCE EN 10029 – Class B
PRODUCT Hot rolled plate softened and descaled 1D Finish **2**		**1**

ITEM	NO	DIMENSIONS (MM)	CAST NO	PLATE NOS
1	3	10 x 2000 x 6000	ABC 9999	4529 4530 4531
2	2	20 x 2000 x 6000	ABC 9999	4532 4533
				3

CHEMICAL COMPOSITION **4**

	C	Si	Mn	P	S	Cr	Ni	Mo	N
Min	0.000	0.00	0.00	0.000	0.000	22.00	4.50	3.00	0.14
Max	0.030	1.00	2.00	0.030	0.015	23.00	6.50	3.50	0.20
ABC 9999	0.017	0.50	1.45	0.019	0.002	22.45	5.43	3.25	0.18

MECHANICAL PROPERTIES **5**

TENSILE TEST TO EN 10002-1 HARDNESS ROCKWELL C

ITEM	DIR	LOC	TEMP	Rp0.2	Rm	A5	A50	HRC
MIN				460	655	25	25	
MAX					840			25
1	T	TOP	20	585	803	32	35	22
1	T	BOT	20	580	795	31	34	23
2	T	TOP	20	570	785	35	36	21
2	T	BOT	20	573	783	34	35	21

CHARPY IMPACT TOUGHNESS TO EN 10045-1 **6**

ITEM	DIR	LOC	TEMP	TEST 1	TEST 2	TEST 3	AVERAGE
MIN			20				85
MAX							
MIN			-50				50
MAX							
1	T	TOP	20	180	185	175	180
1	T	BOT	20	183	190	174	182
2	T	TOP	20	178	189	190	186
2	T	BOT	20	175	196	179	183
1	T	TOP	-50	80	85	75	80
1	T	BOT	-50	83	90	74	82
2	T	TOP	-50	78	89	90	86
2	T	BOT	-50	75	96	79	83

INTERNATIONAL STAINLESS STEEL COMPANY
NEWTOWN
SOMEWHERE
TELEPHONE – 09999 5679989

QUALITY MANAGER

JOHN SMITH

Section 1
This shows the standards and specifications which the material has to meet. In this case, the European EN and the US ASTM standards for general (EN 10088-2 and ASTM A240) and pressure vessel steels (EN 10028-7 and ASME SA 240) are being asked for.

Section 2
A brief description of the products.

Section 3
Dimensions and identity nos, plate nos coil nos, lot nos, etc.

Section 4
The target and actual chemical compositions. This test cert only shows elements which are specified in the grade. Other test certificates may show other residual elements like Cu Sn.

Section 5
The tensile test at room temperature. In this case, the Rm is quite close to the upper limit. Remember the apple picture – 803 apples to break the wire or 80 bags of sugar. This section also shows the hardness test. HRC is the US Rockwell C test. It is normal to specify maximum hardness levels rather than a range.

Section 6
Impact properties are not often measured for austenitic stainless steels. For duplex steels it is normal because they exhibit the ductile-brittle transition. In this case the values are well above the required minimum. Note also the test carried out at minus 50 deg C to ensure acceptable properties in some arduous applications.

Chapter 11 - The Corrosion of Stainless Steels

Corrosion is the bane of our industrial world. The World Corrosion Organisation has estimated that the annual cost of corrosion worldwide is $1800 billion. This corresponds to about 3% of world GDP. According to NACE (the US National Association of Corrosion Engineers). This cost includes:

- Design, manufacturing and construction

Material selection, corrosion resisting alloys to replace carbon and alloy steels, corrosion allowances, use of coatings and paint, corrosion inhibiting chemicals.

- Management

Inspection, maintenance, repairs, replacement, inventory of back-up components, down-time.

The corrosion problem arises from the fact that nature is constantly trying to return metals to the form in which they were originally dug out of the ground. All it needs is air and water to start this process. Without constant intervention, our modern metallic world would crumble into rust and other corrosion products.

Stainless steel has become one of the major factors in combatting corrosion since 1913. For applications in normal atmospheric conditions, stainless steel is effectively inert. For example, a set of stainless steel cutlery will last indefinitely in a normal kitchen.

For stainless steel, corrosion is a subject that most of us wish didn't exist. After all, we have chosen stainless steel to be just that– "stainless". So why is it that it can corrode?

Stainless steel is "stain – less" not "stain – impossible".

Although the passive film is sufficient for most environments, there are harmful chemicals that can break down its defences and lead to corrosion. In urban and coastal locations, the air contains contaminants which are damaging to the passive film. The chemical industry has a whole range of aggressive products which will test the most highly alloyed stainless steel.

Corrosion is quite a difficult subject and we can only scratch the surface in a guide like this. However, it is worth remembering that corrosion involves some kind of electrical cell rather like a battery. Essentially, most metals tend to try to return to their original state in which they are found in the earth, often as an oxide. The oxygen in air or dissolved in water tends to combine with metals to produce a metal oxide.

The main corrosion types are:

- General Corrosion
- Pitting Corrosion
- Crevice Corrosion
- Galvanic Corrosion
- Stress Corrosion Cracking (SCC)
- Microbially Induced Corrosion (MIC)

General Corrosion or Uniform Corrosion

This is a bit like corrosion of carbon steel as the passive film is broken down by the chemical in question and the whole surface is corroded uniformly. This kind of corrosion is usually caused by a strong acid like sulphuric or hydrochloric. Hydrochloric acid in particular is very potent against stainless steels.

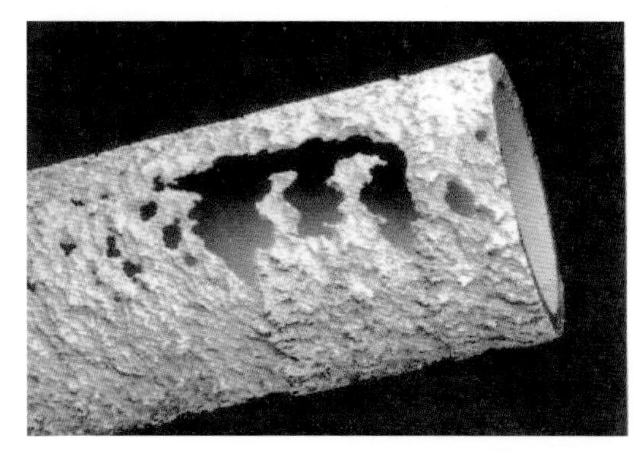

This pipe has been corroded so badly that it has perforated in places.

This is probably due to the wrong grade being chosen or perhaps a wrong assessment of the severity of the conditions.

The resistance to acid corrosion is represented in diagrams similar to the one below. This example is for sulphuric acid.

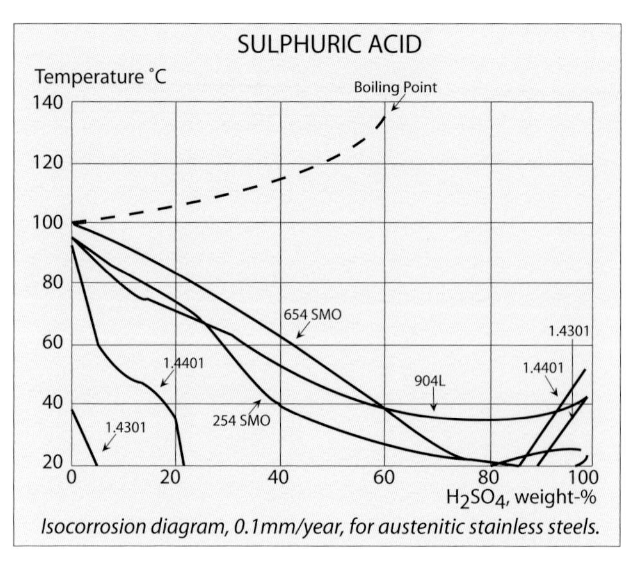

Isocorrosion diagram, 0.1mm/year, for austenitic stainless steels.

For this type of corrosion, a metal loss of 0.1mm/year is regarded as an acceptable limit. From the diagram, the following conclusions can be drawn:

- 1.4301 (304) has very limited use for sulphuric acid service, up to about 5% at 20°C but rapidly reducing to 0% at 40°C.
- 1.4401 (316) has a much better performance than type 1.4301 (304), about 20% at 20°C, only slightly reducing up to about 40°C before tapering off.
- Both these steels are acceptable for very concentrated acids at the right hand end of the diagram.
- The grade 904L (1.4539) was specifically developed for sulphuric acid service and can be used across the whole concentration range.

Pitting Corrosion

This type of corrosion is perhaps the most typical of stainless steels. Although the passive film protects the surface of the steel in most cases, it can be attacked by some chemicals. The most common of these is the chloride ion. This occurs in ordinary salt (sodium chloride). Sources of salt in real applications include salt spray in marine applications and road salt in construction.

The chloride ion is able to penetrate the oxide film and set up a "corrosion cell" which then drives further corrosion.

Pitting can be quite dangerous because it can result in complete localised perforation of the material with quite low metal loss as in this example.

This diagram shows how pitting corrosion occurs. The chloride ion penetrates the passive film and sets up a differential between the inside of the pit and the outside of the pit where the passive film is still doing its job of protecting the surface.

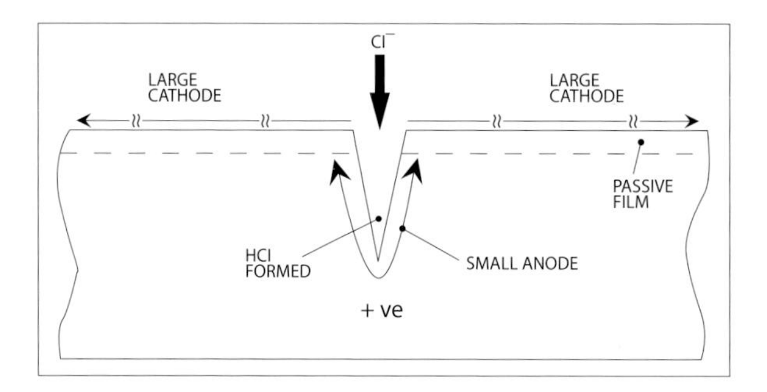

Pitting corrosion resistance is improved by increasing the amount of chromium, molybdenum and nitrogen in the steel. This explains why type 316 (1.4401) with 2% Mo is used in marine atmospheric locations. Mo is about 3 times better than Cr in combatting pitting corrosion.

The susceptibility to pitting corrosion increases with:

- Increased chloride content
- Increased temperature
- Increased acidity (low pH)

Therefore for applications like desalination or other seawater applications, it is often necessary for higher grades than 316 to be used. These include duplex steels such as 1.4462 (2205), 1.4501 (Zeron 100), 1.4410 (2507) and the higher alloy austenitic types such as 1.4539 (904L), 1.4547 (254SMO) and 1.4529 (1925hMo).

The resistance of stainless steel to the initiation of pitting corrosion is measured in a number of ways.

Pitting Resistance Equivalent Number (PREN)

Chromium, molybdenum and nitrogen are the most important elements in combatting pitting corrosion. It has been found that the following formula is useful for comparing the pitting corrosion resistance of stainless steels:

$$PREN = \%Cr + 3.3 \times \%Mo + 16 \times \%N$$

The factor for Nitrogen is open to some debate and values as high as 30 have been proposed.

Some steels use tungsten (W) to improve the pitting resistance. For these steels a modified formula is used:

$$PREN = Cr + 3.3(Mo + 0.5W) + 16N$$

The table below shows the PREN for a series of common stainless steels:

	Type	Cr	Mo	N	PREN
Ferritics					
1.4003	Nirosta 4003/F12N	10.5-12.5	NS	0.030 max	10.5-13.0
1.4016	430	16.0-18.0	NS	NS	16.0
1.4113	434	16.0-18.0	0.9-1.4	NS	19.0-22.6
1.4521	444	17.0-20.0	NS	0.030 max	23.0-28.7
Austenitics					
1.4301	304	17.0-19.5	NS	0.11 max	17.0-20.8
1.4311	304LN	17.0-19.5	NS	0.12-0.22	18.9-23.0
1.4401	316	16.5-18.5	2.0-2.5	0.11 max	23.1-28.5
1.4406	316LN	16.5-18.5	2.0-2.5	0.12-0.22	25.0-30.3
1.4539	904L	19.0-21.0	4.0-5.0	0.15 max	32.2-39.9
1.4547	254SMO	19.5-20.5	6.0-7.0	0.18-0.25	42.2-47.6
1.4529	1925hMo	19.0-21.0	6.0-7.0	0.15-0.25	41.2-48.1
Duplex					
1.4362	SAF 2304	22.0-24.0	0.1-0.6	0.05-0.20	23.1-29.2
1.4462	SAF 2205	21.0-23.0	2.5-3.5	0.10-0.22	30.8-38.1
1.4410	SAF 2507	24.0-26.0	3.0-4.0	0.24-0.35	37.7-46.5
1.4501	Zeron 100	24.0-26.0	3.0-4.0	0.2-0.3	37.1-44.0

Critical Pitting Temperature

The PREN has been found to be related to a measure of pitting resistance called the Critical Pitting Temperature. This is the temperature at which a steel grade will start to show pitting in a particular solution.

The solutions used are usually quite strong. A typical test is the ASTM G 48 Method E test which uses a 6% Ferric Chloride with 1% Hydrochloric Acid solution.

The graph below shows the relationship between the critical pitting temperature (CPT) in this test with the PREN values for a range of stainless steel grades.

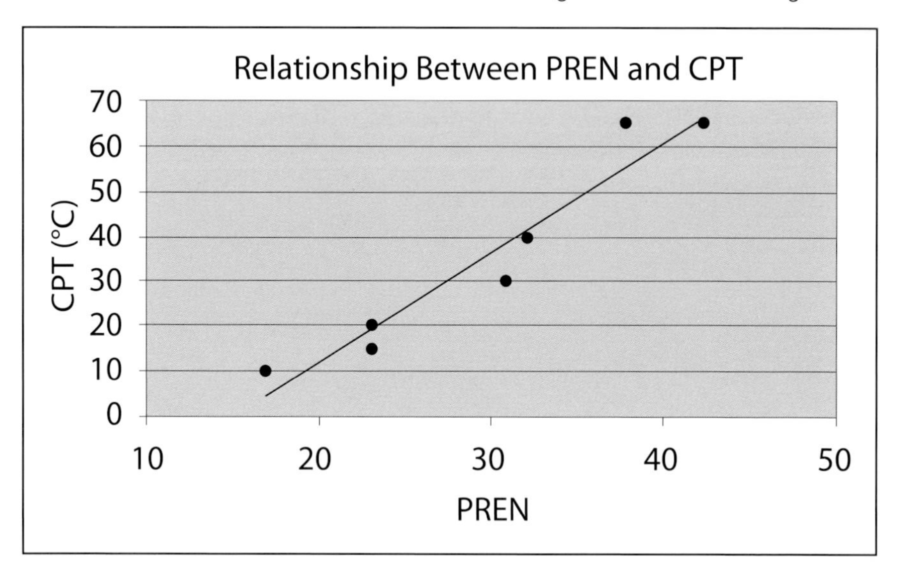

Whilst nickel does not contribute to the resistance to initiation of pitting corrosion, it does have a beneficial effect on the subsequent propagation of pitting.

Crevice Corrosion

This form of corrosion can occur where a stainless steel face is brought into direct contact with another metallic or hard inflexible face. A typical situation would be a bolted flange in a tube system. Inside this crevice, there is insufficient oxygen for the passive film to form properly and there is a differential set up between the inside and outside of the crevice. If the fluid flowing along the pipe is sufficiently aggressive this can cause corrosion as shown in the picture. One solution to the problem is to use some kind of flexible sealant which completely seals off the crevice from the fluid as in the diagram on the right. Threads of bolts can also be a cause of crevice corrosion.

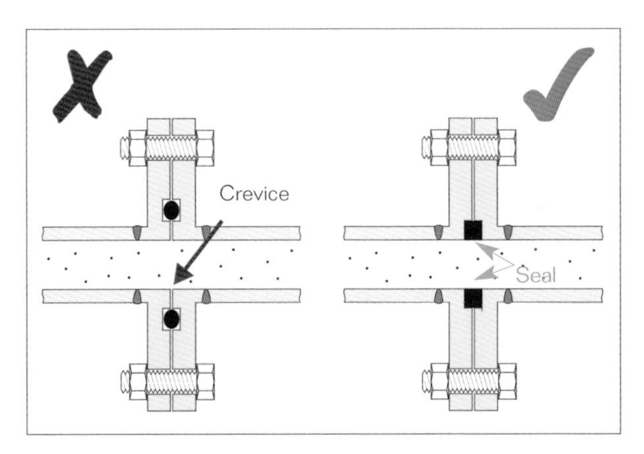

Crevice corrosion resistance is improved by the same elements as pitting corrosion – Cr, Mo and N.

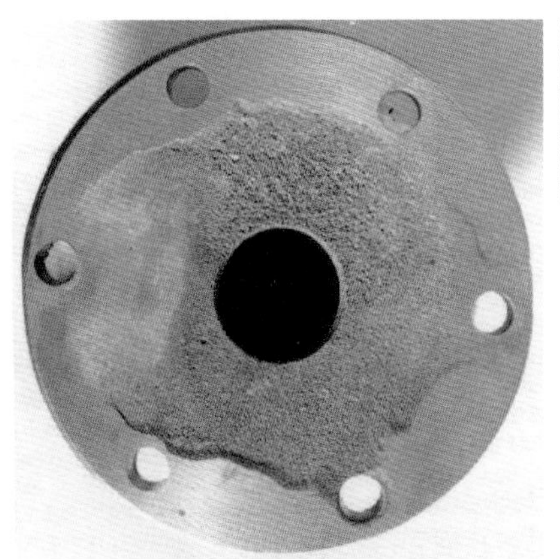

The resistance to crevice corrosion is measured in a similar way to pitting corrosion by finding the Critical Crevice Temperature in aggressive solutions. A typical test is the ASTM G 48 Method E test. In this case, a series of crevices is formed by fastening 2 inflexible crenellated washers to the surfaces of the steel sample. The samples are then immersed in the same acidified ferric chloride solution as for the Critical Pitting Test.The temperature at which crevice corrosion first occurs is the Critical Crevice Temperature (CCT).

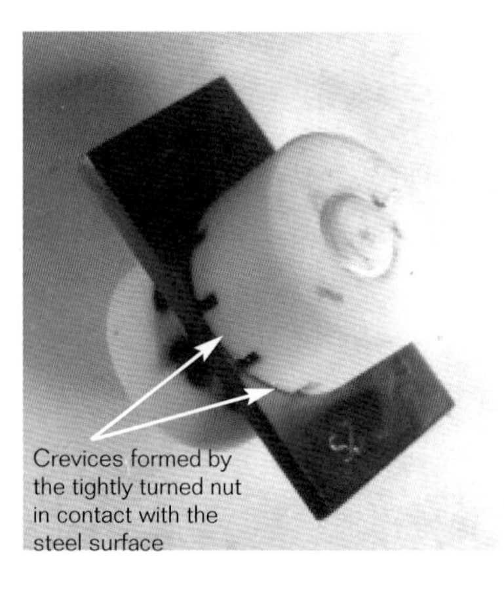

The CCT is usually lower than the CPT for each grade as shown in the table:

Grade	*CPT(°C)*	*CCT(°C)*
1.4307	10	< 0
1.4401	20	< 0
1.4462	30	20
1.4539	40	10
1.4547	65	35
1.4410	65	35

Galvanic or Bimetallic Corrosion

This form of corrosion can occur when two different metals are in contact with each other and joined by a fluid. The factors affecting whether or not corrosion actually occur are:

- The difference between the two metals in the galvanic series
- The relative surface area of the two metals
- The severity of the joining fluid

The galvanic series is essentially a ranking of how reactive a metal is. A typical series is shown below:

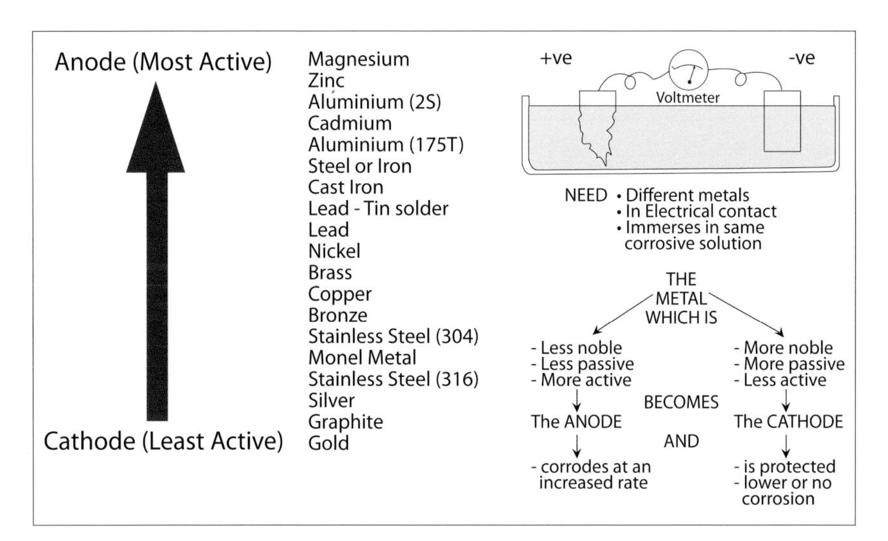

The most common possibilities for galvanic corrosion involving stainless steels are with aluminium, carbon steel or galvanised steel (zinc coated). Even with these metals, galvanic corrosion will not occur if the surface area of the more active metal is larger than the stainless steel. Therefore it is generally acceptable to use stainless steel fasteners to join carbon steel, galvanised steel and aluminium but not to use carbon steel, galvanised steel or aluminium fasteners through stainless steel. Ordinary water such as rainwater can be sufficient to trigger galvanic corrosion, so it is important to consider metal to metal contacts in external architectural applications and also in humid internal applications like swimming pools. Process piping systems provide another major sector where dissimilar metals can be in contact. It is good design practice to isolate dissimilar metals wherever possible. Sometimes painting over the join is a simple and effective method. In some situations it may be necessary to adopt a more sophisticated approach.

Metals like copper, brass and nickel are too close to stainless steel in the series to be a high risk.

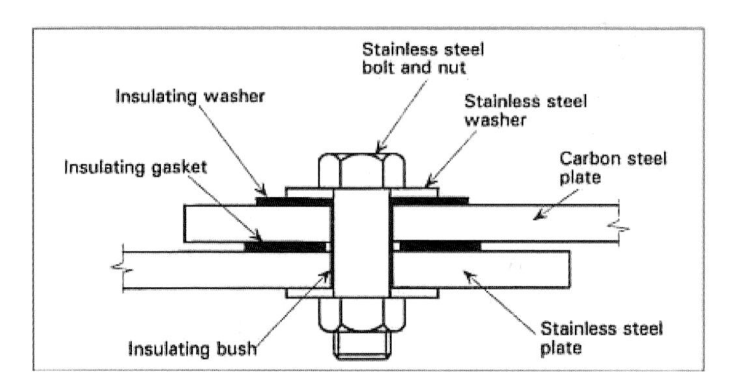

Intergranular Corrosion

When austenitic stainless steels were being developed, it was found that corrosion often occurred near welds. This phenomenon was called "weld decay". It was found that the cause of the problem was due to the formation of Chromium Carbide in the heat affected zone of the weld. At high carbon levels (0.08%) chromium carbide forms quite rapidly between about 900 and 500°C as the weld cools down. This chromium carbide effectively takes chromium away from the steel matrix and reduces the content below that which is required to form the passive film. The depletion of chromium occurs near the grain boundaries. Hence, intergranular (between the grains) corrosion can therefore occur quite readily.

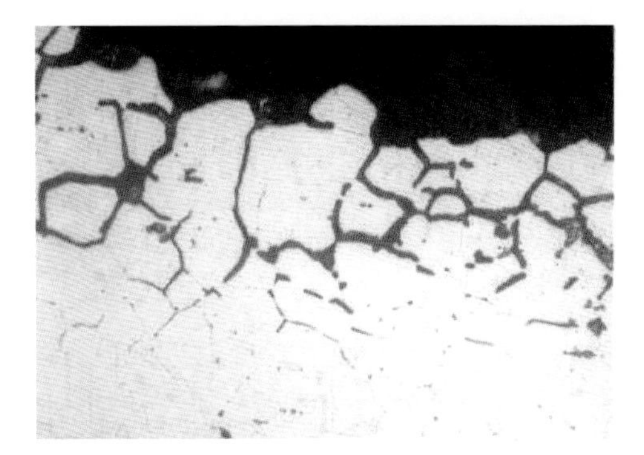

In extreme cases, grains can actually fall out of the matrix of the steel as shown in the picture.

There are two ways by which this phenomenon can be avoided:

- Add an element such as titanium (Ti) or niobium (Nb) which combines with the carbon instead of chromium, leaving the chromium in the matrix to do its job of passive film formation. This is why type 1.4541 (321) with Ti and 1.4550 (347) with Nb were developed.

- Lower the carbon to a level where the chromium carbide formation cannot occur during the cooling from welding. This is how grades 1.4307 (304L) and 1.4404 (316L) work. The diagram below shows the effect of carbon level on the formation of chromium carbide.

With 0.08% carbon, chromium carbide will begin to form in less than 1 min. This is well within the timescale of cooling of welds, particularly in thick sections. With 0.03% carbon, carbide formation only starts after *several hours*.

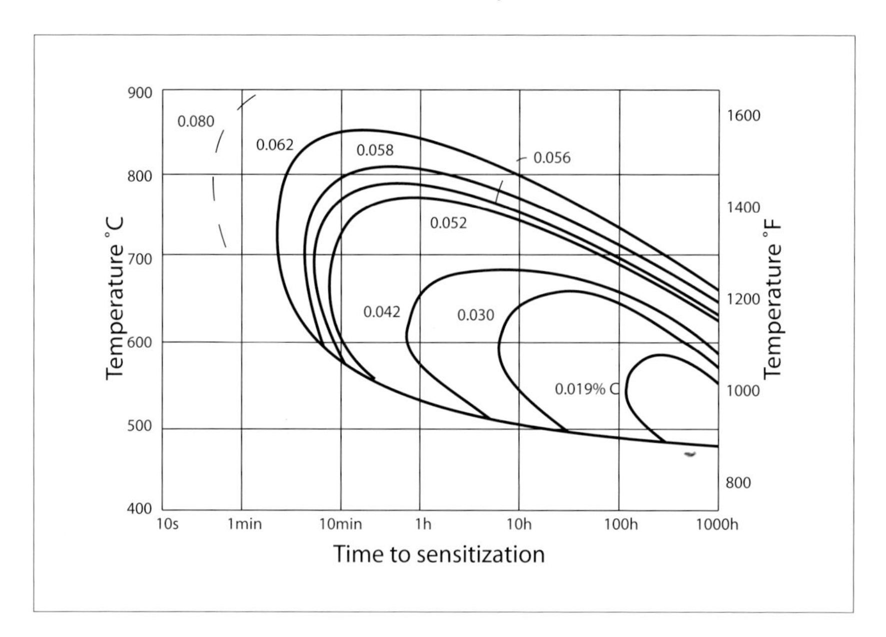

Stress Corrosion Cracking (SCC)

This form of corrosion requires the following conditions:

- Sufficiently high temperature
- Tensile stress
- A corrosive medium

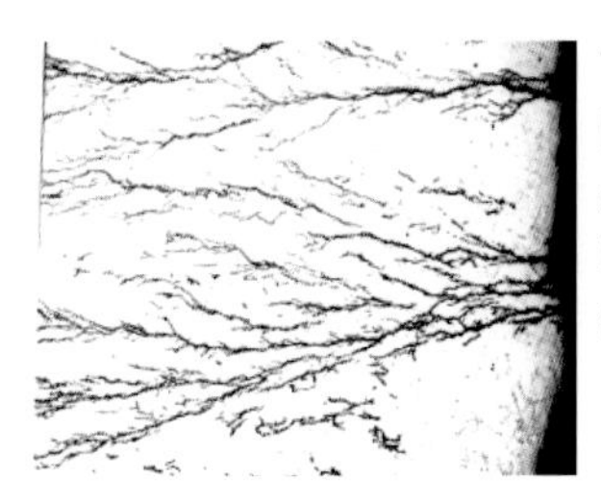

This is what stress corrosion cracking typically looks like. In most situations, SCC does not occur below about 50°C. It is most often found in chemical and food processing which use solutions at high temperatures.

However, it has also been found that the conditions occurring in some swimming pools can lead to SCC at much lower temperatures – 25-30°C. This was discovered after an accident in a Swiss pool in 1985. This involved a concrete ceiling suspended by 1.4301 (304) rods. Unfortunately, the rods gradually failed and the ceiling fell into the pool killing a number of people. Following this, intensive research was carried out to find the cause and solution to the problem.

The cause was traced to the build up of compounds called chloramines which are formed from the chlorine used to disinfect the pool and bodily fluids. These chloramines evaporate and condense on metal surfaces. In areas such as the roof space it is difficult to clean the surfaces and so these chemicals can become highly concentrated. Even at the relatively low temperatures found in swimming pools it was found that the concentration of chloramine was sufficient to trigger SCC.

The solution to the problem is to use stainless steels for structural applications which are much more resistant to SCC than the standard austenitic stainless steels. These include:

- Duplex stainless steels e.g. 1.4462, 1.4410, 1.4501
- Highly alloyed austenitic stainless steels e.g. 1.4529, 1.4539, 1.4547, 1.4565

The problem is now well understood and designers know that grades of stainless steel are able to meet the demands of the environment.

In general, ferritic stainless steels are also resistant to stress corrosion cracking. Grades such as 1.4509 and 1.4510 are used for hot water tanks in preference to 304.

As with pitting and crevice corrosion, there are laboratory tests which compare the SCC resistance of the various grades of stainless steel. A typical test is the drop evaporation test. This involves dripping a sodium chloride solution onto a heated sample, thereby giving a high concentration of chloride on the surface. The samples are stressed up to the 0.2%PS of the material. The minimum stress at which SCC occurs is measured for each material. Typical values for a range of steels are shown in the table below:

Steel Grade	Ratio of Actual Stress to 0.2% Proof Stress (%)	
1.4301 (304)	< 10	The higher the ratio, the more resistant is the grade.
1.4432 (316L)	< 10	
1.4462 (2205)	30	
1.4362 (2304)	40	
1.4410 (2507)	60	
1.4539 (904L)	60	
1.4547 (254SMO)	80	

Although stress corrosion cracking often involves the chloride ion, there is a form of stress corrosion cracking which is common in the oil and gas industry. This occurs with so-called "sour wells" where hydrogen sulphide is present.

Microbially Induced Corrosion

Many process solutions contain bacteria. In some conditions, notably stagnant ones, bacteria can build up a "slime" on the surface of the steel. Underneath this slime, oxygen can become reduced and therefore does not allow the passive film to form. This is similar to crevice corrosion.

Chapter 12 - The High Temperature Properties of Stainless Steel

The passive film on stainless steel is also able to resist gaseous atmospheres at high temperatures. Atmospheres can be oxidising, sulphidising, reducing and carburising. In contrast to corrosion, these processes are essentially dry.

The elements which are used to improve resistance to gaseous atmospheres are somewhat different to those which are used to improve corrosion resistance:

- Chromium)
- Silicon) All these elements strengthen the passive film.
- Aluminium)
- Nickel – This has the effect of making the passive film more adherent to the surface of the steel and making it less likely to flake off.
- Rare earth metals – Elements like cerium are used in small amounts to improve the passive film.

The absence of molybdenum from this list is notable. In general, molybdenum does not provide any benefit in gaseous atmospheres and in some situations is detrimental.

With the correct combinations of elements, steels can be formulated which can be used at temperatures as high as 1150°C. The table below shows the maximum temperature of use for a number of common heat resisting grades:

Grade	Maximum Temperature (According to EN 10095)
1.4833 (309)	1000
1.4845 (310)	1050
1.4835 (253MA)	1150
1.4841 (314)	1150
1.4828 (306)	1000
1.4818 (153MA)	1050

This is only a guideline as the final choice depends on the composition of the gaseous atmosphere.

Some applications require both high temperature resistance and resistance to corrosion. Flue gas chimneys are typical. When gases condense they can form quite aggressive liquids, often acidic. The choice of grade is something of a compromise. 1.4301 (304) or 1.4401 (316) are typically used with 1.4539 (904L) being used where the condensate is particularly aggressive.

Mechanical Properties at High Temperatures

When stainless steels are used at elevated temperatures, it is also important to understand the effect of temperature on the strength.

The graphs below show the fall in 0.2% PS and UTS for a range of common grades.

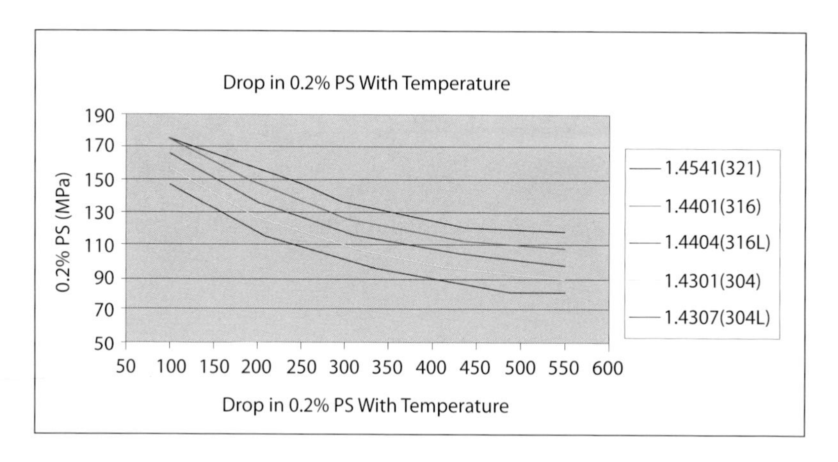

Drop in 0.2% PS With Temperature

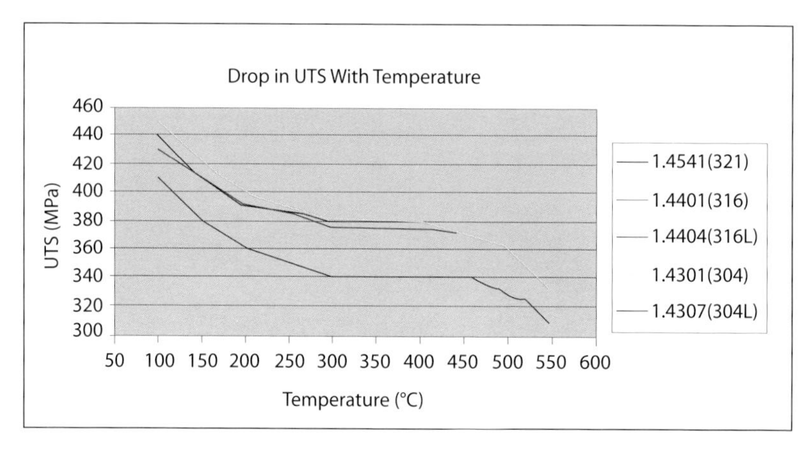

Of course, stainless steels are used at temperatures well above the limit of the graphs. At these temperatures, a phenomenon known as creep starts to come into play.

- Creep is time dependent deformation. This means that a sample will continue to stretch over time without increasing the load.

The resistance to creep is measured by:

- The maximum stress which can be applied to give a certain strain at a specified temperature in a particular time OR

- The maximum stress which can be applied to fracture (rupture) a specimen at a specified temperature in a particular time

The times can be very long; 10,000 hours (1 year) and 100,000 hours (11.5 years) are typical of laboratory tests.

The importance of creep can be judged by comparing the simple UTS with the stress to rupture. Grade 1.4541(321) is a common grade which is used for creep resistance. At 550°C, the minimum UTS is 330 MPa. This compares to only 223 MPa for the 10,000 hour creep rupture strength and 170 MPa for the 100,000 hour creep rupture strength.

Creep values are not usually reported on test certificates for release testing purposes. The times involved are far too long to do this. Many results have been collected over the years to give minimum safe values for designers to use.

Chapter 13 - Physical Properties of Stainless Steels

Physical properties usually refer to thermal expansion, thermal conductivity and electrical resistivity.

Thermal expansion

When a substance is heated it usually increases in size. Austenitic stainless steels expand more than carbon steels and ferritic or martensitic stainless steels. The thermal expansion coefficient is the increase in unit length for each increase in temperature of 1°C. The thermal expansion coefficient also increases with temperature. The table below shows the values given in EN 10088-1 for some common grades.

	Coefficient of Thermal Expansion (10^{-6} per °C) between 20°C and Temperature Shown				
Grade	**100°C**	**200°C**	**300°C**	**400°C**	**500°C**
1.4003 (3CR12)	10.4	10.8	11.2	11.6	11.9
1.4512 (409)	10.5	11.0	11.5	12.0	12.0
1.4000 (410S)	10.5	11.0	11.5	12.0	12.0
1.4016 (430)	10.0	10.0	10.5	10.5	11.0
1.4028 (420)	10.5	11.0	11.5	12.0	N/A
1.4057 (431)	10.0	10.5	10.5	10.5	N/A
1.4301 (304)	16.0	16.5	17.0	17.5	18.0
1.4307 (304L)	16.0	16.5	17.0	18.0	18.0
1.4401 (316)	16.0	16.5	17.0	17.5	18.0
1.4404 (316L)	16.0	16.5	17.0	17.5	18.0
1.4541 (321)	16.0	16.5	17.0	17.5	18.0
1.4833 (309)	N/A	16.0	16.7	17.5	17.8
1.4845 (310)	N/A	15.5	16.2	17.0	17.3
1.4539 (904L)	15.8	16.1	16.5	16.9	17.3
1.4462 (2205)	13.0	13.5	14.0	N/A	N/A

To visualise what these small numbers mean consider the following example:

How much would a metre long rod expand when heated from 20°C to 200°C in grade 1.4301 (304) and grade 1.4016 (430)?

1.4301 (304)
From the table the thermal expansion coefficient is 16.5×10^{-6} per °C. (This can also be expressed as 0.0000165 per °C). As the increase in temperature is 180°C, the total expansion is $180 \times 16.5 \times 10^{-6}$ metres = 0.00297 metres or about 3 mm.

1.4016 (430)

As the thermal expansion for this grade is only 10.0 x 10^{-6} per °C, the expansion is 180 x 10.0 x 10^{-6} metres = 0.0018 metres or 1.8mm.

For comparison, carbon steel has an expansion coefficient of about the same as ferritic stainless steels. Aluminium is about 23 x 10^{-6} per °C.

Thermal conductivity

This is the ability of a material to allow heat to pass through it. Metals are usually good conductors. However, austenitic stainless steels have a lower thermal conductivity than carbon steels and ferritic or martensitic stainless steels. The lower thermal conductivity of austenitic stainless steels explains why stainless steel saucepans have an aluminium or copper bottom to get an even temperature for cooking the food.

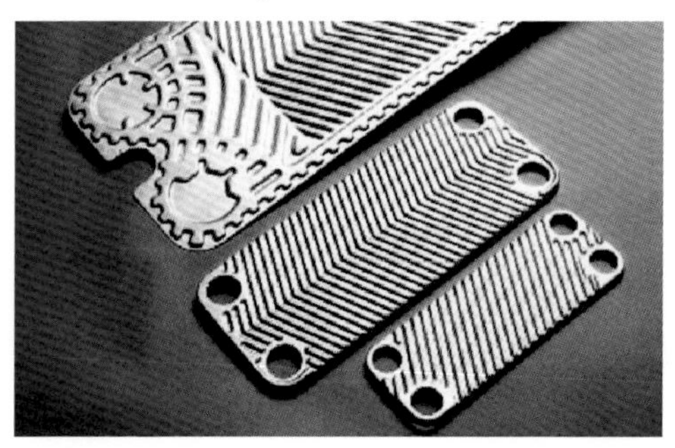

Electrical resistivity

This is the ability of a material to resist electric current passing through it. Metals are usually good conductors. However, austenitic stainless steels have a higher electrical resistivity than carbon steels and ferritic or martensitic stainless steels.

The table below shows the values of these two properties given in EN 10088-1 for some common grades:

Grade	Thermal Conductivity at 20°C (W per m per °C)	Electrical resistivity at 20°C (Ohm mm^2 per m)
1.4003 (3CR12)	25	0.60
1.4512 (409)	25	0.60
1.4000 (410S)	30	0.60
1.4016 (430)	25	0.60
1.4028 (420)	30	0.65
1.4057 (431)	25	0.70
1.4301 (304)	15	0.73
1.4307 (304L)	15	0.73
1.4401 (316)	15	0.75
1.4404 (316L)	15	0.75
1.4541 (321)	15	0.73
1.4833 (309)	15	0.80
1.4845 (310)	15	0.85
1.4539 (904L)	12	1.00
Carbon steel	~35	~0.2
Aluminium	> 200	~0.03

Practical Implications of Physical Properties

The most important effect of these properties, particular for the austenitic stainless steels, is for welding. The high thermal expansion and low thermal conductivity means that distortion is more likely to occur than with carbon steels. It is important to make sure that welds are securely restrained and that heat is conducted away from the weld areas to avoid such distortion.

Tack welding can also be used to minimise distortion. Tack welds are temporary means of holding components in proper alignment until the final and definitive welding is completed.

Chapter 14 - The Ingredients of Stainless Steel

The following table lists the different elements used in stainless steels and their effects on properties:

Element	Symbol	Approximate Content Range %	Properties Affected	Grades Showing Effect of Element
Chromium	Cr	10.5 – 30	Cr is the essential ingredient of stainless steel. It improves: Pitting corrosion resistance Crevice corrosion resistance General corrosion resistance High temperature oxidation resistance	All stainless steels
Nickel	Ni	0 – 37	Ni is used to make the austenitic structure stable at normal temperatures. The austenitic structure improves the formability and weldability of stainless steels. Low Ni for stretch forming High Ni for deep drawing High Ni austenitic stainless steels are very resistant to stress corrosion cracking. Ni also improves the high temperature oxidation.	 1.4301 8% Ni 1.4301 9% Ni 1.4303 (305) 1.4539 (904L) 1.4547 1.4529 1.4845 (310)
Molybdenum	Mo	0 – 6	Mo improves: Pitting corrosion resistance Crevice corrosion resistance General corrosion resistance It may be detrimental in some high temperature applications It gives a small increase in strength	1.4521 (444) 1.4401 (316) 1.4404 (316L) 1.4462 (2205) 1.4539 (904L) 1.4547 1.4529

Carbon	C	0.01 – 1	C is a very powerful element in stainless steels. In austenitic stainless steels, low C (< 0.030%) gives improved weldability by reducing carbide precipitation and therefore susceptibility to intergranular corrosion. However, low C reduces strength particularly at high temperature. A minimum C of 0.04% is specified to give a higher strength. The range of C in martensitic steels is very large up to 1%. Increasing levels of C lead to increasing strength but reducing ductility and toughness.	1.4307 (304L) 1.4404 (316L) 1.4948 (304H) 1.4028 (420) 1.4031 1.4112 (440B) 1.4125 (440C)
Silicon	Si	Up to 2.5	Found in all stainless steels as a residual element. Principally used to improve high temperature oxidation resistance.	1.4828 (306)
Manganese	Mn	Up to 11	Mn is an alternative to Ni to produce the austenitic structure in the so-called 200 series steels. Although cheaper than Ni the mechanical properties, weldability and formability are generally inferior to Ni bearing steels. Mn is therefore not a direct substitute for Ni. Mn steels tend to be more work hardening than Ni steels.	1.4372 (201) 1.4373 (202)
Nitrogen	N	Up to 0.6	N is used in austenitic and duplex stainless steels to improve: Pitting corrosion resistance Crevice corrosion resistance Strength N is added to reduce magnetic permeability at low temperature. High N reduces hot workability and makes a steel more difficult to produce.	1.4406 (316LN) 1.4439 (317LMN) 1.4462 (2205) 1.4410 (2507) 1.4311 (304LN) 1.4406 (316LN)

Titanium	Ti	Up to 2.3	Used to combine with C to avoid intergranular corrosion in steels like 1.4541 (321) particularly where higher strength is required compared to 1.4307 (304L).	1.4541 (321)
			In ferritic stainless steels used to improve corrosion resistance often in conjunction with Nb.	1.4509 (441) 1.4510 (439)
Niobium	Nb (Cb from columbium in US)	Up to 1.5	Used to combine with C to avoid intergranular corrosion in steels like 1.4550 (347) particularly where higher strength is required compared to 1.4307 (304L). Not common in Europe except in aerospace applications.	1.4550 (347)
			In ferritic stainless steels used to improve corrosion resistance often in conjunction with Ti.	1.4509 (441)
Aluminium	Al	Up to 2	Al is used in ferritic stainless steels to improve the high temperature oxidation resistance.	1.4742 1.4762
			In precipitation hardening steels like 1.4568 (17-7 PH), the hardening effect is due to Al-rich particles in the matrix of the steel.	1.4568 (17-7 PH)
Copper	Cu	Up to 5	Cu is used to improve resistance to reducing acids like sulphuric acid, notably in 1.4539 (904L).	1.4539 (904L)
			Also used in "cold heading" grades for fastener production due to reduction in work hardening.	1.4567 (304Cu)
			In precipitation hardening steels like 1.4542 (17-4 PH), the hardening effect is due to Cu-rich particles in the matrix of the steel.	1.4542 (17-4 PH) 1.4594 (FV 520B)

Tungsten	W (from wolfram, another name for tungsten)	Up to 3.5	Similar to Mo. Used in duplex stainless steels up to 1.0% to improve pitting corrosion resistance. Used in creep resistant steels up to 3.5%	1.4501 1.4905 1.4945 1.4971
Cerium	Ce	Up to 0.1	Used to improve high temperature oxidation resistance.	1.4818 1.4835
Vanadium	V	Up to 0.85	Particularly used in high temperature creep resistant alloys in conjunction with Cr and Mo.	1.4980 (A286)
Sulphur	S	Up to 0.35	S of 0.25/0.35% is used in free machining stainless steels. These give markedly inferior corrosion resistance compared to normal S levels of < 0.030. Controlled S levels of 0.015/0.030% in conjunction with Calcium treatment are used to improve machinability whilst not impairing corrosion resistance.	1.4305 (303)
Calcium	Ca	Up to 0.03%	Used in conjunction with controlled S to control inclusions in stainless steel to improve machinability.	
Boron	B	Up to 1.2%	Low levels of boron 0.0015/0.0050% are used to improve hot workability and creep strength High levels of boron > 0.5% are added to type 304 to absorb neutrons in the handling of spent nuclear fuel	1.4919 (316H) 1.4941 (321H) No EN grade, proprietary grades

Cobalt	Co	Up to 21%	Used in creep resistant alloys Very low Co levels < 0.10% are required in standard austenitic steels like 304 for nuclear applications	1.4911 1.4971
Zirconium	Zr	Up to 0.15%	In ferritic stainless steels used to improve corrosion resistance by combining with carbon	1.4590

Chapter 15 - The Manufacture of Stainless Steel

Flat Products

The process route shown is for a standard austenitic stainless steel.

All stainless steel starts with melting in an electric arc furnace (EAF).
Typical capacity 100-150 tonnes.
Scrap accounts for 60-70% of a new melt or cast.
Temperatures in the arc furnace are around 1500°C.

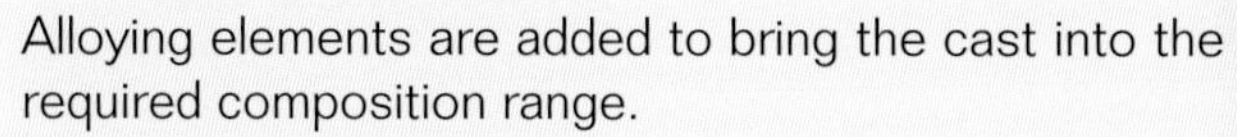

The molten steel is transferred to a converter usually an AOD vessel. (Argon oxygen decarburisation). Here the carbon is blown out using oxygen. The invention of this process in 1954 and its subsequent development led to a step change in the cost of production of stainless steels due to the efficient reduction in carbon levels particularly for the low carbon "L" grades.

Alloying elements are added to bring the cast into the required composition range.

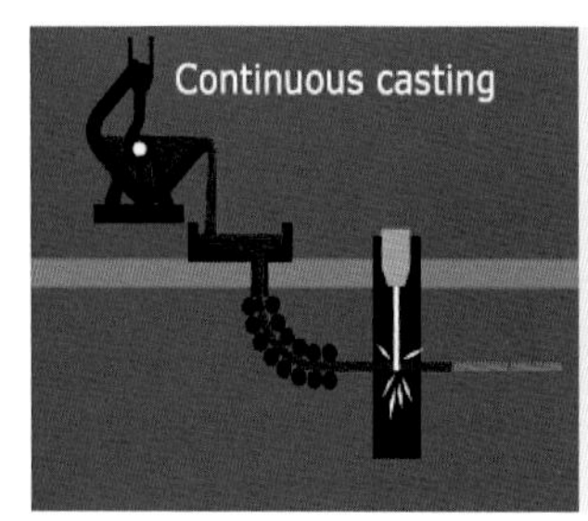

The molten steel is then poured vertically into a mould. As the steel solidifies it is formed by a series of rolls into a horizontal slab typically 1000 – 2000 mm wide by 100 - 300 mm thick. The slab is then flame cut into lengths suitable for hot rolling. Slab grinding may also take place to make the resulting hot rolled surface better.

The invention of the continuous casting (concast) process was another major step in reducing the cost of production.

The cast structure is very coarse and not suitable for many applications in this state. Some form of working, often by rolling, is required to break down the cast structure into a finer, more useful "wrought" structure.

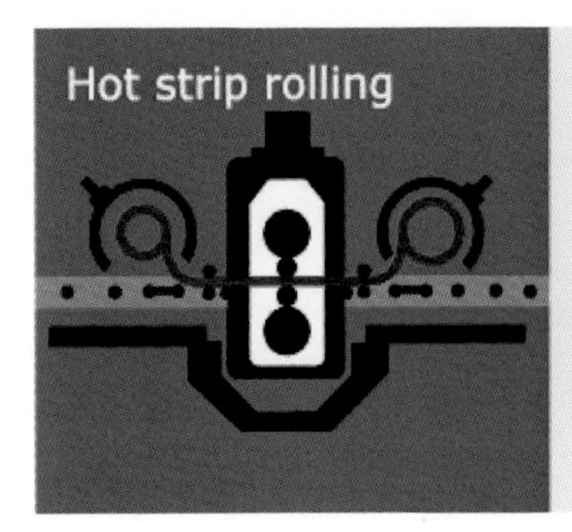

The slab is then rolled after heating to about 1250°C. In a Steckel Mill, the coil is kept hot in furnaces at either side of the rolls. At this stage the width of the coil is the same as the original slab. The thickness is reduced to between 13 and 3 mm dependent on the next stage in the process. At this stage the product is called Black Hot Band.

Quarto plate is produced as individual pieces and remains flat throughout the process. This product can be up to about 150mm thick.

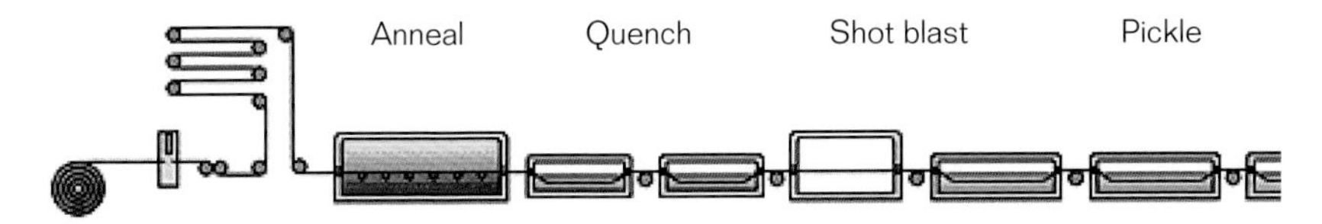

The coil is then sent down a continuous line which combines a number of processes;

Anneal – Heat to approximately 1050°C. This is a solution anneal. This softens the steel so that it can be further processed by cold rolling or in its end use.

Quench – The strip is rapidly cooled to avoid any harmful phases coming out of solution.

Shot blast and pickle – In hot rolling a high-temperature oxide scale is formed. This needs to be removed by mechanical and chemical means to give a surface which is capable of forming the passive layer.

At this stage the coil of steel is in a condition which can actually be sold. The surface is defined as 1D in EN 10088-2. This is equivalent to the older designations:

HRAP – Hot Rolled Annealed and Pickled
HRSD – Hot Rolled Shot Blasted and Descaled

It has a non-reflective white/light grey surface. It is also quite rough having an Ra value of typically 5 micron.

The product may also be referred to as CPP – Continuously Produced Plate. If it is to be processed further into cold rolled coil it can also be referred to as White Hot Band.

Quarto plate is produced in a similar way except that the pieces of plant are not necessarily in a continuous line.

Cold rolling of the white hot band is done by a Sendzimir mill known as a "Z" mill. This uses a cluster of rolls which concentrates the rolling force onto a small surface area. The rolls are highly polished to give a smooth surface to the stainless steel.

Cold rolled thicknesses range from about 0.3 mm to 8 mm in wide coil. In narrow coil, typically at 300 mm wide, much thinner gauges down to less than 0.1mm can be produced.

As the strip is reduced in thickness it work hardens and becomes more difficult to roll. For the thinnest material, rolling may take place in two or more stages with intermediate solution annealing (softening). This explains why thin material is more expensive than thick material.

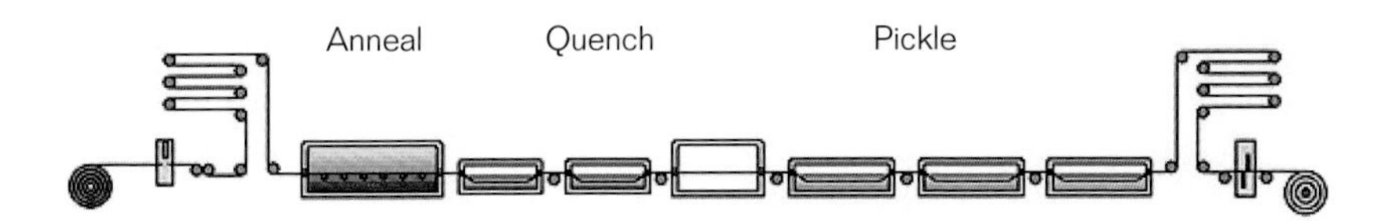

After cold rolling the strip must be given a final anneal. This is a similar process to the initial process except that now the surface of the steel is very smooth. Therefore, the aggressive shot blasting is not required. The final pickle is only needed to remove the relatively light oxide scale from the annealing operation.

At this stage, the material is classed as 2D. In practice, most cold rolled material goes through a pinch passing (skin passing) operation to produce the most common 2B surface.

Finishing

This is a very light rolling which slightly brightens the surface and improves the shape of the coil for further processing.

For applications requiring a very high degree of flatness, a tension levelling process can be applied.

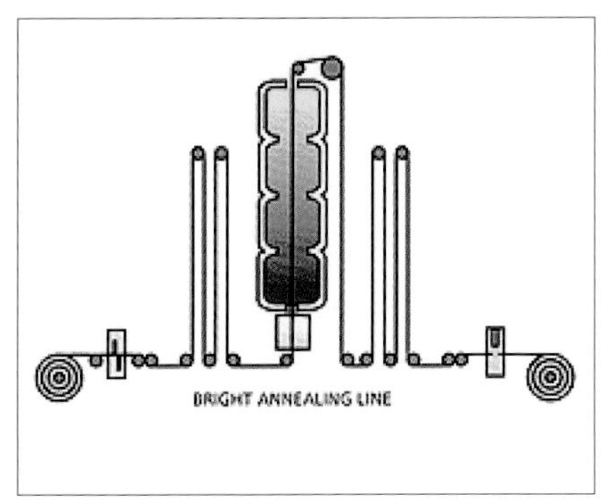

An alternative to the 2B surface is bright annealed. This is now officially known as 2R in EN 10088-2. However, the traditional BA designation will continue to be used for a long time.

Bright annealing takes place in an inert atmosphere of nitrogen and hydrogen so that the bright smooth cold rolled surface is preserved.

Coils are produced in standard widths of typically 1000 mm, 1250 mm and 1500 mm. 2000 mm wide is also available. Precision strip with tighter thickness tolerances is produced in narrower standard widths.

Slitting is used to produce narrow coils.

Sheets are cut to length from a coil.

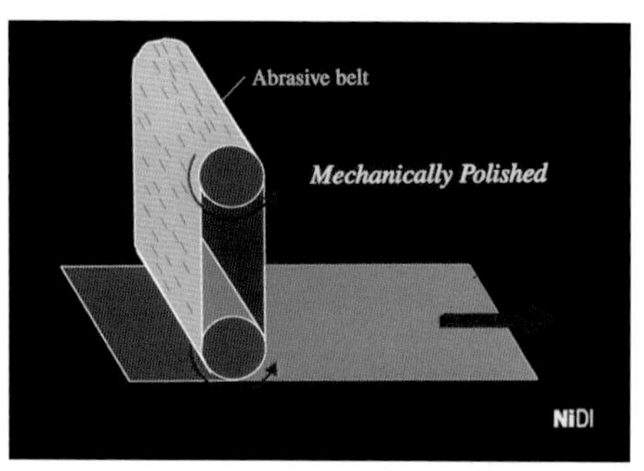

Mechanical polishing using abrasives such as alumina or silicon carbide is a common method for producing a uniform directional polish on a sheet.

Long Products – Bars, rod and wire

Essentially the melting and refining processes are the same as for flat products. From the casting process onwards there are distinct differences.

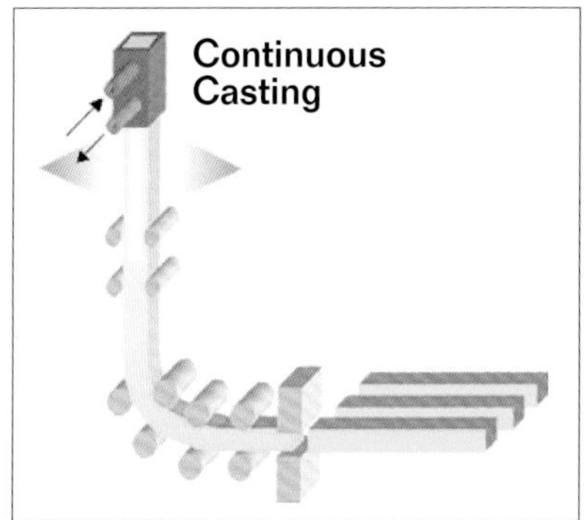

For long products the starting point is a bloom or a billet.
These are usually square/rectangular in cross section. There is no clear distinction between a bloom and billet, except that a bloom is larger than a billet.

Typical bloom size 200 x 200 mm
Typical billet size is approx 150 x 150 mm

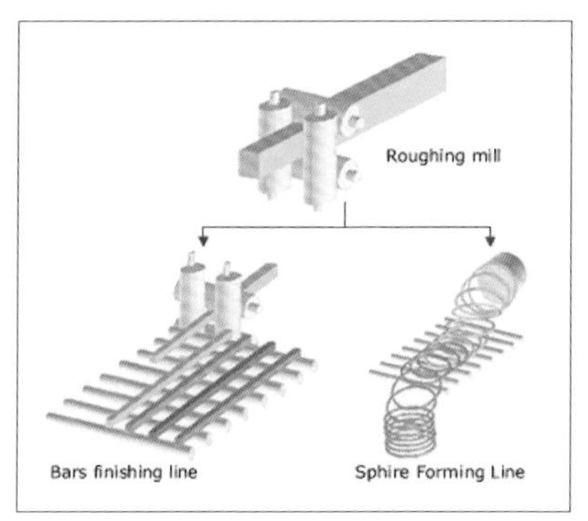

Blooms/billets can then be further rolled into bar or rod in coil

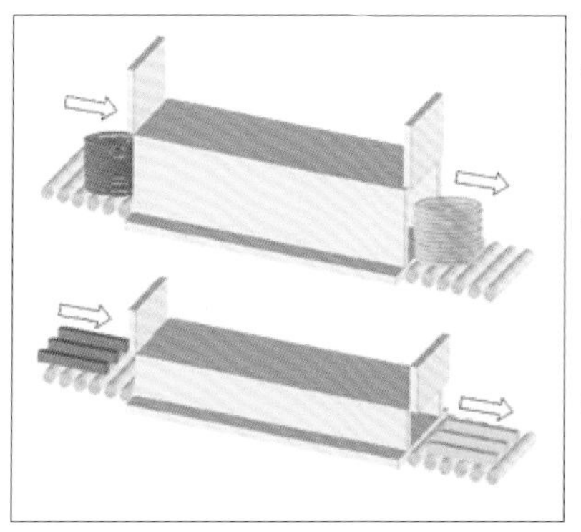

These are then annealed in a similar way to flat products.

Annealing of coil

Annealing of bar

Followed by pickling

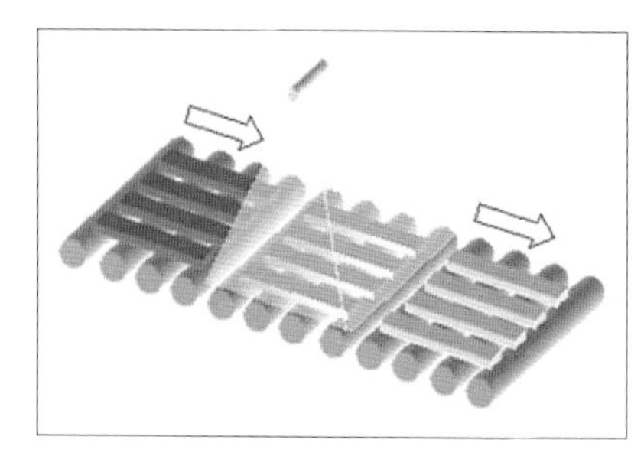

Pickling of bar

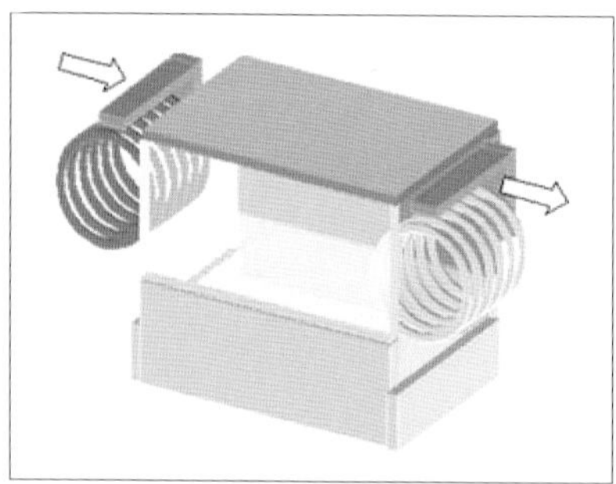

Pickling of coil

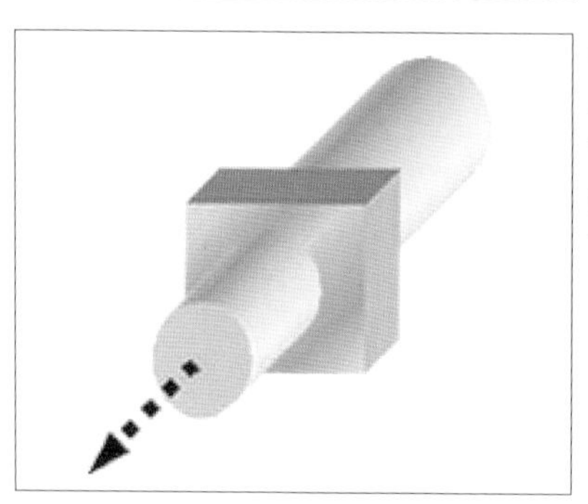

Cold drawing of bar/coil to give smooth surface, good dimensional tolerance. Also work hardens the product. In general, long products have higher tensile properties than flat products due to this process.

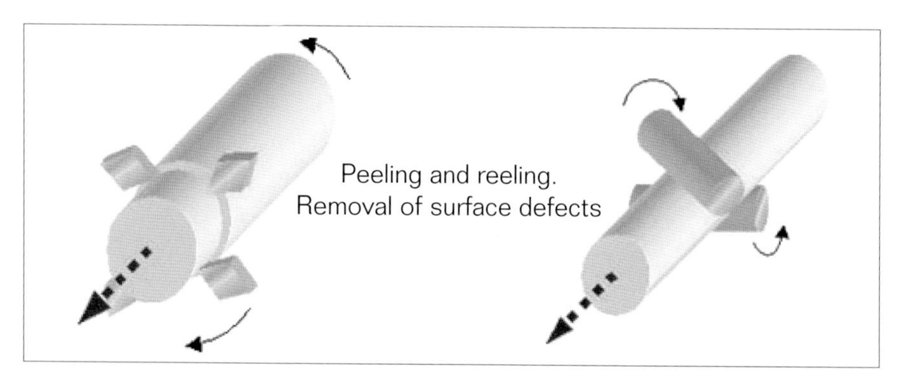

Peeling and reeling.
Removal of surface defects

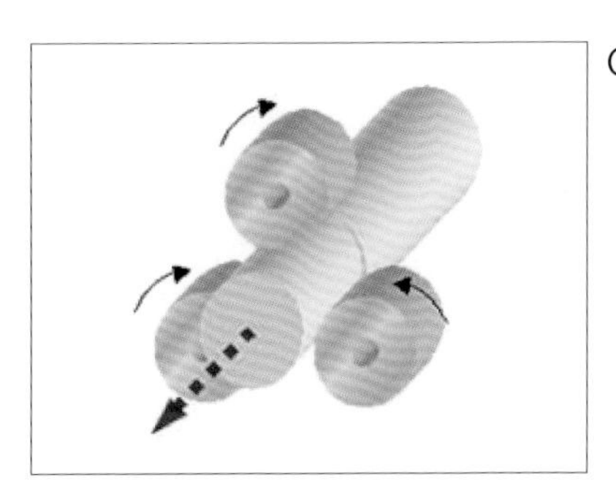

Grinding

Tube Products – Continuously welded from coil

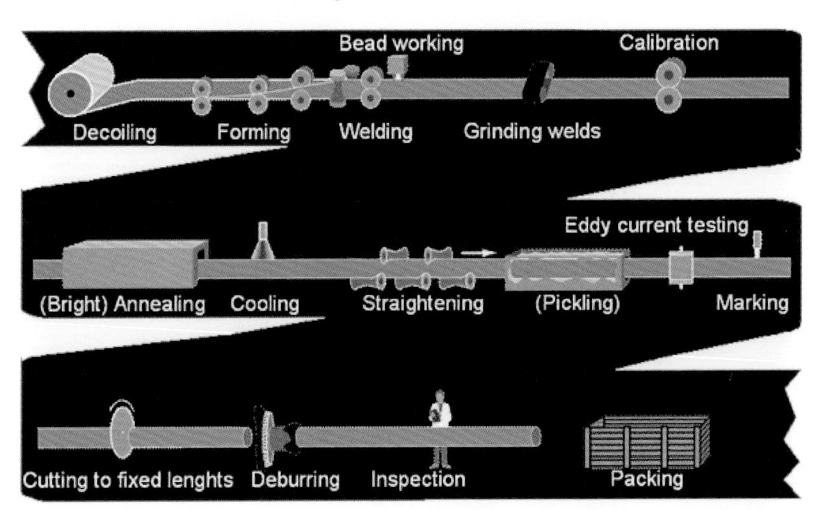

Tube and Pipe – welded from individual sheets or plates

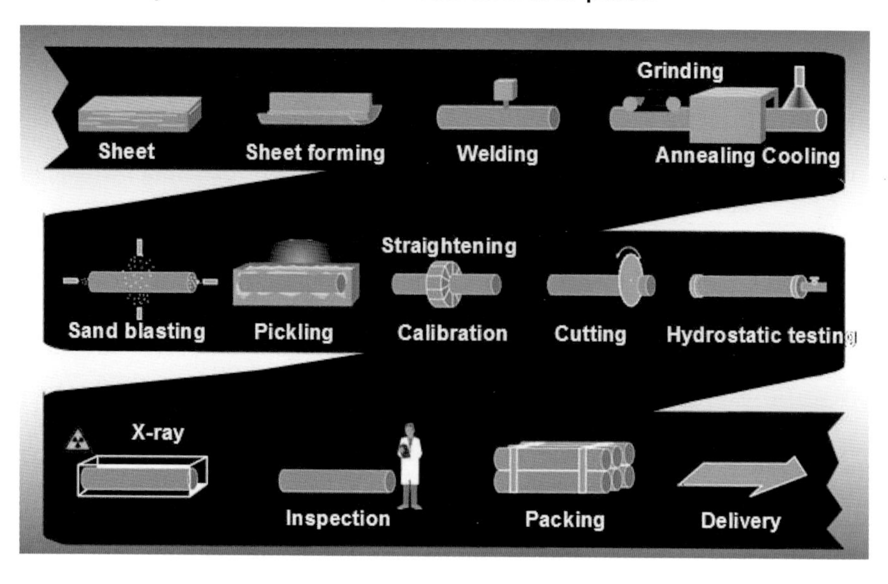

Tube - seamless

Seamless tube is often produced by extrusion as shown in the diagram:

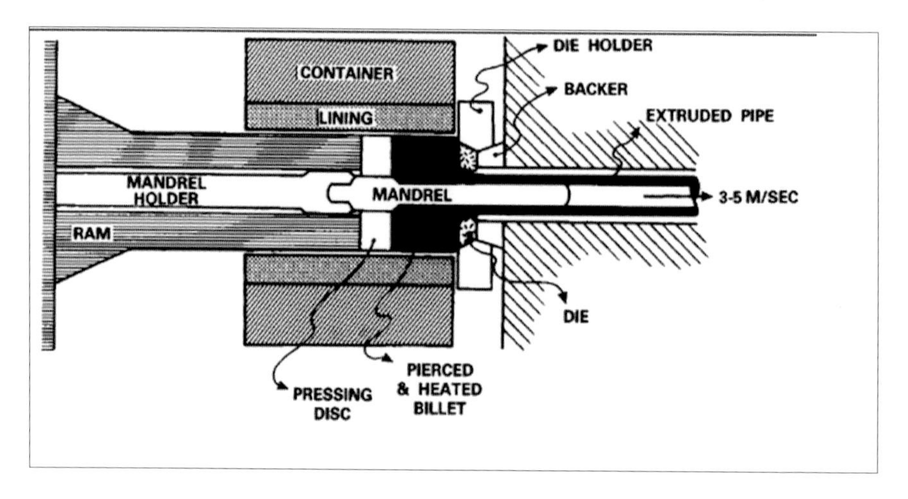

Chapter 16 - Surface Finishes on Stainless Steel

One of the great attractions of stainless steel is that it can be produced in many surface finishes. This gives architects and designers a wide range of possibilities when designing in stainless steel:

Bright annealed **2B**

Bright polished

Brushed, satin polish, dull polish

Patterned finishes

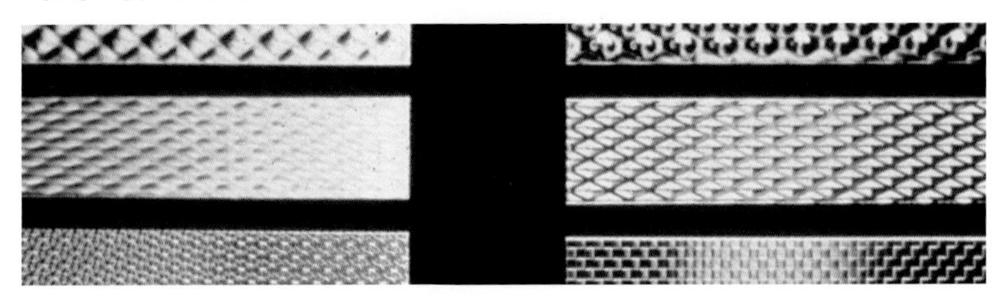

Coloured finishes

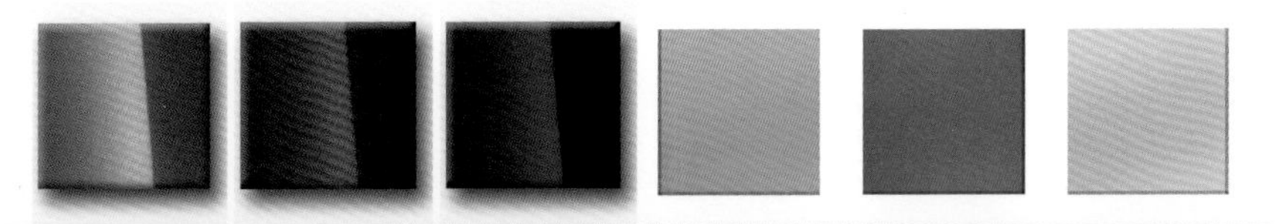

Surface Finish and Corrosion Resistance

One of the most important aspects of surface finish is its effect on the corrosion resistance. There are numerous examples where designers and specifiers have not understood this relationship and consequently a poor corrosion performance has followed.

The European Standard EN 10088-2 defines surface finishes summarised in the following table.

Surface Finish Code	Description	Typical Surface Roughness Ra (micron)
Mill finishes		
1D	Hot rolled, heat treated, pickled. The most common hot rolled finish. A non reflective, rough surface. Not normally used for decorative applications.	4-7
2B	Cold rolled, heat treated, pickled, pinch passed. The most common cold rolled mill finish. Dull grey slightly reflective finish. Can be used in this condition or frequently is the starting point for a wide range of polished finishes.	0.1-0.5
2D	Cold rolled, heat treated, pickled.	0.4-1.0
2F	Cold rolled, heat treated, skin passed on roughened rolls. Produces a dull matt finish.	
2H	Work hardened by rolling to give enhanced strength level. Various ranges of tensile or 0.2% proof strength are given in EN 10088-2 up to 1300 MPa and 1100 MPa respectively dependent on grade.	
2Q	Cold rolled hardened and tempered. Applies to martensitic steels which respond to this kind of heat treatment.	
2R	Cold rolled and bright annealed, still commonly known as BA. A bright reflective finish. Can be used in this condition or as the starting point for polishing or other surface treatment processes e.g. colouring.	05-0.1

In the following codes "1" refers to hot rolled being the starting point and "2" as cold rolled.		
Special Finishes		
1G or 2G	Ground. Relatively coarse surface. Unidirectional. Grade of polishing grit or surface roughness can be specified.	
1J or 2J	Brushed or dull polished. Smoother than 1G/2G. Grade of polishing grit or surface roughness can be specified.	0.2-1.0
1K or 2K	Satin polish. Similar to 1J/2J but with maximum specified Ra value of 0.5 micron. Usually achieved with SiC polishing belts. Alumina belts are strongly discouraged for this finish as this will have detrimental effect on corrosion resistance. Recommended for external architectural and coastal environments where bright polish (1P/2P) is not acceptable.	less than0.5
1P/2P	Bright polished. Non-directional, reflective. Can specify maximum surface roughness. The best surface for corrosion resistance.	less than 0.1
2L	Coloured by chemical process to thicken the passive layer and produce interference colours. A wide range of colours is possible.	
1M/2M	Patterned. One surface flat.	
1S/2S	Surface coated e.g. with tin = Terne coating.	
2W	Corrugated. Similar to patterned but both surfaces are affected.	

Problems often arise where generic descriptions are used instead of the "official" definitions.

Typical descriptions are:

Dull polish
Satin polish
Brushed

These terms on their own are effectively meaningless. Any reference of this type on drawings, order/enquiry documents and specifications should always be clarified.

The important thing to understand is that the surface roughness on a micro scale affects the corrosion resistance of the steel. Surface roughness is measured by using the Ra value in micron (a micron is one thousandth of a millimetre). Consider the examples below. These are both samples of 1.4401 (316) stainless steel:

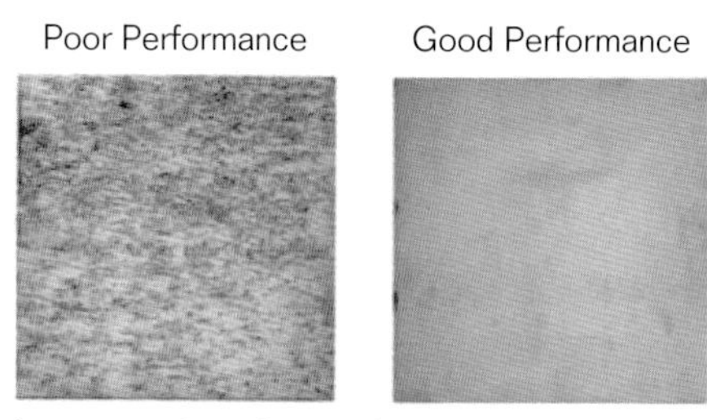

Appearance of stainless steel surface after accelerated salt spray test.

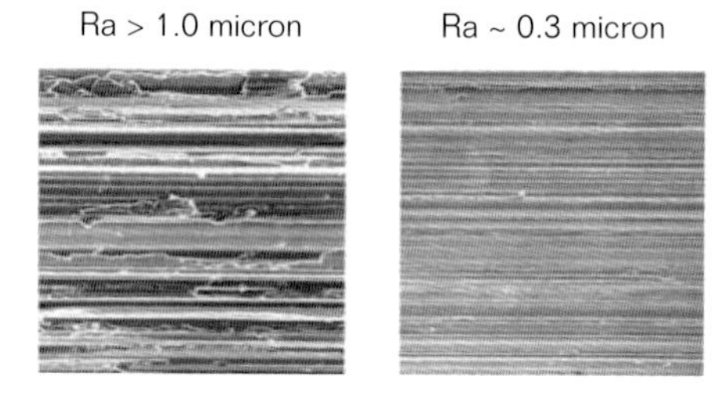

Appearance of samples on a microscopic scale. Both samples have been ground using 240 grit polishing belts. The key point is that the rough sample has been polished with an alumina (Al_2O_3) belt and the smooth sample with a silicon carbide (SiC) belt.

The difference in corrosion resistance is illustrated by the different behaviour in an accelerated salt spray test. This difference is reflected in real environments, particularly marine and urban.

Rough

Smooth

Three factors are at work here:

- Rougher surfaces allow corrosive material to "stick" more easily in the first place.
- Having got there, rougher surfaces are less effective in allowing natural washing by rainwater to remove the corrosive material.
- The rough surface in itself is more likely to provide sites for crevice corrosion.

It follows that very smooth surfaces like bright annealed or bright polished are the least prone to surface staining. These have Ra values of about 0.1 micron.

However, bright reflective surfaces are not always acceptable for architectural applications for aesthetic reasons. It has been found that a polished surface with a maximum surface roughness of 0.5 micron is necessary to impart a satisfactory corrosion resistance for external applications. This is embodied in the 2K surface finish defined in EN 10088-2 for flat products.

The need for such a minimum standard can be illustrated by an example from the Stainless Steel Advisory Service.

An enquiry was received asking for advice on the corrosion of 1.4401 (316) stainless steel "satin polished" flat bar. The enquirer was apparently a long established supplier of the product and had "never had a problem before". It turned out that the surface roughness on the product was about 3.0 micron. Not surprisingly the material corroded quite badly in a marine environment.

From a corrosion point of view, the 2B mill surface is actually very good, having typical surface roughness values of 0.1-0.5 micron. However, aesthetically, the surface is often too variable to use in its raw state.

Surface Finish and Hygiene

The food processing and catering industries are among the largest consumers of stainless steel. The combination of corrosion resistance and ease of cleaning make it a material of choice for these applications.

Much research has been carried out on the "cleanability" of stainless steel. As might be expected, the surface roughness is an important factor. On rough surfaces, bacteria can "hide" in the micro-crevices. The European Hygienic Engineering and Design Group (EHEDG) has determined that a maximum surface roughness of 0.8 micron will give a good cleanability. In practice, the need for a 0.5 micron surface for corrosion resistance means that surfaces will be well under this threshold.

Chapter 17 - Fabrication of Stainless Steels

Stainless steel can be fabricated in similar ways to carbon steel by:

- Cutting
- Pressing
- Bending
- Welding
- Etc

However, there are some very important differences between the two materials that need to be taken into account:

- Optimisation of yield – stainless steel is an expensive material compared to carbon steel so the cutting of the parent materials needs to be carefully planned to avoid unnecessary waste.
- Segregation of carbon and stainless steels – contamination of stainless steel surfaces can occur from carbon steel in the form of grinding dust, welding sparks or transfer from tools which have been used on both materials. This can occur both in the fabrication shop and on site. Such contamination quickly corrodes and ruins the appearance of the stainless steel.

Stainless steel contaminated by contact with carbon steel

- Properties such as thermal conductivity and thermal expansion lead to increased risk of distortion which needs to be taken into account. The use of restraining jigs and tack welds are techniques used to prevent distortion.

- After welding it is important to deal with the heat tint that occurs due to the high temperature oxidation of the surface. This oxide is not like the passive film and is vulnerable to corrosion. It is important to remove this oxide.

The diagram below shows how the corrosion resistance of the welded area varies with the type of treatment given.

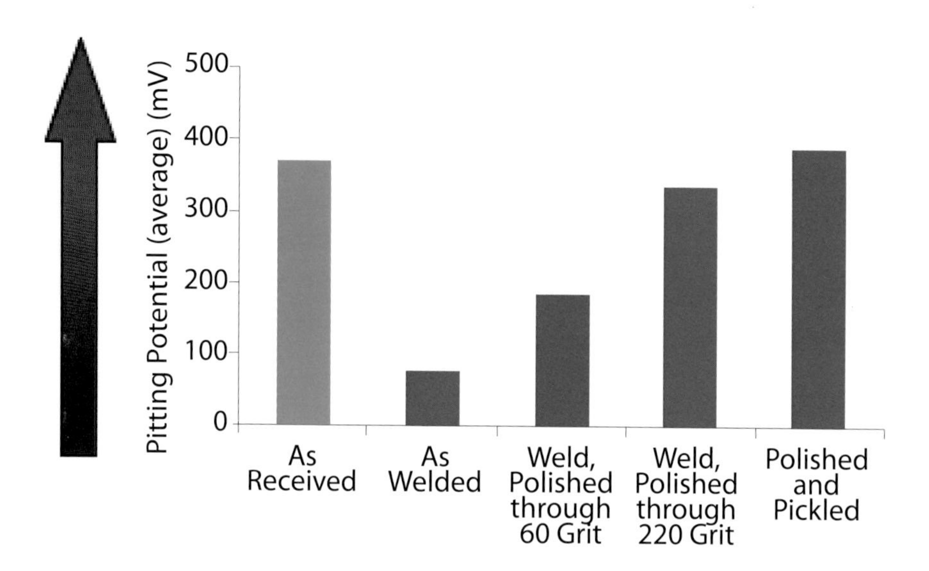

Pickling of the weld is the best way to restore the original corrosion resistance of the steel. Ideally all welds should be pickled.

- As the surface finish is often the main reason for using stainless steel, it should be protected for as long as possible during the fabrication process. Plastic films are normally used for this purpose.

- The use of strong chemicals such as masonry or brick cleaner should be avoided. These can contain hydrochloric acid which is highly corrosive towards stainless steel.

Chapter 18 - Material Selection

Unfortunately, material selection is a rather imprecise science. You would have thought that, with all the experience obtained since the invention of stainless steel, it would be easy to decide which steel to use for a particular application. Not so. At least it keeps metallurgists in gainful employment!

Material selection often ends up being a compromise between the optimum technical solution and two other practical considerations:

- Material cost.
- Material availability.

It may be thought that steel X is the best grade for the job but if no-one can supply it in a reasonable time or if it costs £20000 per tonne it is unlikely to be used.

The degree of risk is also important to consider. If the consequence of failure is serious e.g. high rectification costs, long down time or potential injury, proper care should be taken in material selection.

With this in mind, here is a list of questions which should be addressed when trying to choose a suitable material:

1. What is the corrosive environment? – Atmospheric, water, concentration of particular chemicals, chloride content, presence of acid etc. It is particularly difficult when there is a combination of chemicals, some of which may never have been tested with stainless steels.

2. What is the temperature of operation? – High temperatures usually accelerate corrosion rates and therefore indicate a higher grade. Low temperatures will require a tough austenitic steel.

3. What strength is required? – Higher strength can be obtained from the austenitic, duplex, martensitic and PH steels. Other processes such as welding and forming often influence which of these is most suitable. For example, high strength austenitic steels produced by work hardening would not be suitable where welding was necessary as the process would soften the steel.

4. What welding will be carried out? - Austenitic steels are generally more weldable than the other types. Ferritic steels are weldable in thin sections. Duplex steels require more care than austenitic steels but are now regarded as fully weldable. Martensitic and PH grades are less weldable.

5. What degree of forming is required to make the component? – Austenitic steels are the most formable of all the types being able to undergo a high degree of deep drawing or stretch forming. Ferritic steels are better than austenitics in deep drawing but not as good in stretch forming. Duplex, martensitic and PH grades are not particularly formable.

6. What product form is required? – Not all grades are available in all product forms and sizes, for example sheet, bar, tube. In general, the austenitic steels are available in all product forms over a wide range of dimensions. Ferritics are more likely to be in sheet form than bar. For martensitic steels, the reverse is true.

7. What are the customer's expectations of the performance of the material? – This is an important consideration often missed in the selection process. Particularly, what are the aesthetic requirements as compared to the structural requirements? Design life is sometimes specified but is very difficult to guarantee.

8. There may also be special requirements such as non-magnetic properties to take into account.

9. It must also be borne in mind that steel type alone is not the only factor in material selection. Surface finish is at least as important in many applications, particularly where there is a strong aesthetic component.

10. Availability. There may be a perfectly correct technical choice of material which cannot be implemented because it is not available in the time required. Some grades may not be available in all product forms, except for the standard austenitic steels like 304 and 316.

11. Cost. Sometimes the correct technical option is not finally chosen on cost grounds alone. However, it is important to assess cost on the correct basis. Many stainless steel applications are shown to be advantageous on a life cycle cost basis rather than initial cost alone.

12. Sometimes it is not cost per se that is the problem but the cost stability. The variability in cost due to alloys can be a factor.

13. Is there any similar application which could give a "steer" towards the correct grade?

14. What is the risk of failure?

15. What are the consequences of failure? If serious damage or fatalities could result from an incorrect choice, material selection will be on the safe side. If consequences are minor, then a little less caution would be required.

The final choice will almost certainly be in the hands of a specialist but their task can be helped by gathering as much information about the above factors. Missing information is sometimes the difference between a successful and unsuccessful application.

Some examples of material selection will help to illustrate the issues:

Example 1 – Domestic Sink

Factors affecting choice:

- Reasonable corrosion resistance. Domestic conditions relatively mild – except bleach.
- Ease of cleaning.
- Ease of manufacture – formability, specifically stretch forming.
- Reasonable cost.

Choice = Type 1.4301 (304) with low Ni for optimum stretch forming properties.

Example 2 – Automotive Exhaust

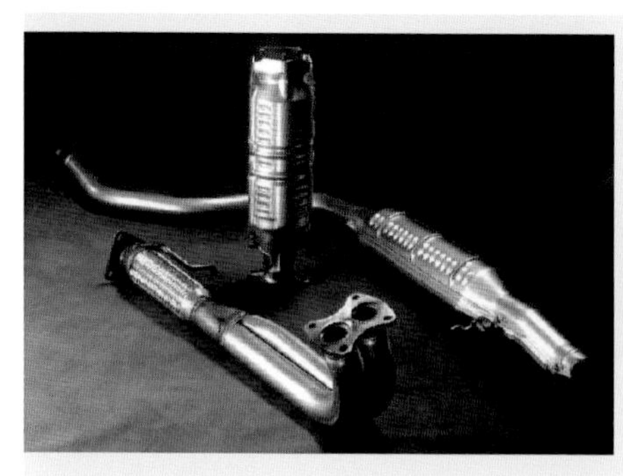

Factors affecting choice:

- Basic corrosion resistance.
- Cosmetic factor low (at least on the invisible part!).
- Meet warranty.
- Ease of manufacture – formability.
- Low cost – auto industry driven by cost.

Choice = Type 1.4512 (409) cheapest ferritic steel with necessary forming characteristics.

Example 3 – Pharmaceutical Vessel .

Factors affecting choice:

- Good corrosion resistance.
- Cosmetic factor high.
- Ease of manufacture – weldability in thick sections.
- Reasonable cost.

Choice = Type 1.4404 (316L) – norm for pharmaceutical industry.

Chapter 19 - Recycling Of Stainless Steel

In common with many other metals, stainless steel is fully recyclable. The Cr, Ni and Mo in stainless steel scrap makes it a very valuable commodity. The stainless steel industry is therefore geared to the large scale recovery of the material.

Also, it is more cost effective to recycle scrap than it is to use new materials.

Recycling of scrap reduces the carbon footprint of stainless steel compared to using new materials.

Here is a typical scrap cycle

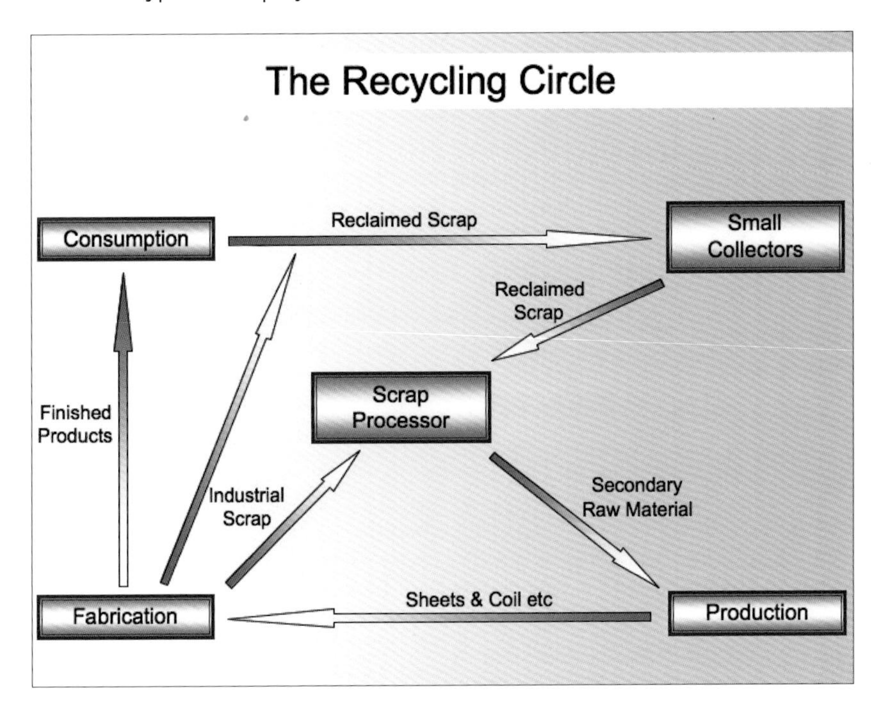

Chapter 20 - The Basics of Stainless Steel

Hopefully, this book has been interesting and will help you in your specific job. But supposing there was only one sheet of paper on which to set out the most important issues concerning stainless steel, what would they be. Here is a list of 10 things that you can photocopy and have pinned to the wall next to your desk. These are the ones that will save most disappointment, time, effort, money and ear-bending by dissatisfied customers:

1. Surface finish is as important as grade in affecting corrosion resistance. "Brushed", "satin polish" are almost meaningless. Surface roughness Ra value is the key factor. Silicon carbide abrasives are better than alumina. "Bright is best".

2. Post weld treatment of stainless steel should be the rule not the exception.

3. Bad design can result in poor performance e.g. partially sheltered will reduce the beneficial effect of natural washing by rain.

4. Stainless steel should be segregated from carbon steel to avoid contamination, particularly when processing or fabricating.

5. Maintenance of stainless steel should be considered in the design process. Stainless steel is not self-cleaning!

6. Stainless steel is not immune to corrosion – select grade for environment avoid chemicals like masonry cleaner.

7. Stainless steel more than just 430 304 and 316.

8. Magnetic properties not always a sure guide in distinguishing grades. Austenitics can become magnetic, particularly cold drawn bar, wire and temper rolled sheet.

9. Don't assume the customer is right! You can be the one to avoid disaster if you ask the awkward question!

10. When to ask for help! (See page 117 for useful sources of information).

Chapter 21 - Common Standards for Stainless Steel

The European system of standards is extensive. One of the principles of the system is that there is a parent standard for each main product form and a number of referenced standards which deal with tolerances, test methods etc. The following table summarises the referenced standards for the standards relating to stainless flat, long and tube products:

Standard	Product Form	Related Standards	Subject
EN 10088-2	Flat	EN 10088-2	Chemical Compositions
		EN 10088-2	Mechanical properties
		EN 10258	Tolerances for narrow cold rolled strip
		EN 10259	Tolerances for wide cold rolled strip/sheet/coil
		EN 10051	Tolerances for hot rolled coil/sheet
		EN 10029	Tolerances for quarto plate
		ISO 9444	Tolerances for hot rolled coil/sheet. (To supersede EN 10051)
		ISO 9445	Tolerances for all cold rolled products (To supersede EN 10258/9)
		ISO 18286	Tolerances for quarto plate (To supersede EN 10029)
		EN 10079	Definition of product forms
		EN 10002	Tensile test methods
		EN 10045	Impact test methods
		EN 10163-2	Surface condition of hot rolled flat products
		EN 10168	Inspection documents
		EN 10204	Types of inspection documents
		EN 10307	Ultrasonic testing of flat products
		EN ISO 377	Test piece sampling
		EN ISO 3651-2	Intergranular corrosion testing
		EN ISO 6506-1	Brinell hardness testing
		EN ISO 6507-1	Vickers hardness testing
		EN ISO 6508-1	Rockwell hardness testing
		EN ISO 14284	Sampling for chemical composition testing

EN 10088-3	Long	EN 10088-3	Chemical Compositions
		EN 10088-3	Mechanical properties
		EN 10058	Tolerances for hot rolled flat bar
		EN 10059	Tolerances for hot rolled square bar
		EN 10060	Tolerances for hot rolled round bar
		EN 10061	Tolerances for hot rolled hexagon bar
		EN 10278	Tolerances on bright steel products
		EN 10017	Tolerances on wire rod
		EN 10218-2	Tolerances on wire
		ISO 286-1	System of limits and fits
		EN 10079	Definition of product forms
		EN 10002	Tensile test methods
		EN 10045	Impact test methods
		EN 10163-3	Surface condition of sections
		EN 10168	Inspection documents
		EN 10204	Types of inspection documents
		EN 10306	Ultrasonic testing of H beams and IPE beams
		EN 10308	Ultrasonic testing of bars
		EN ISO 377	Test piece sampling
		EN ISO 3651-2	Intergranular corrosion testing
		EN ISO 6506-1	Brinell hardness testing
		EN ISO 6507-1	Vickers hardness testing
		EN ISO 6508-1	Rockwell hardness testing
		EN ISO 14284	Sampling for chemical composition testing
EN 10296-2	Welded Tube	EN 10296-2	Chemical Compositions
		EN 10296-2	Mechanical Properties
		EN ISO 1127	Tolerances. However, some tolerances are also covered within EN 10296-2 itself
		EN 10079	Definition of product forms
		EN 10002	Tensile test methods
		EN 10246-2	Eddy current testing for leak tightness

		EN 10246-3	Eddy current testing for detection of imperfections
		EN 10246-7	Peripheral ultrasonic testing for detection of longitudinal imperfections
		EN 10246-8	Weld seam ultrasonic testing for detection of longitudinal imperfections
		EN 10246-9	Weld seam ultrasonic testing of submerged arc welded tubes for detection of longitudinal/transverse imperfections
		EN 10246-10	Weld seam radiographic testing for detection of imperfections
		EN ISO 8491	Bend test for tubes
		EN ISO 8492	Flattening test for tubes
		EN ISO 8493	Drift expanding test for tubes
		EN ISO 8496	Ring tensile test for tubes
		EN ISO 377	Test piece sampling
		EN ISO 3651-2	Intergranular corrosion testing
EN 10297-2	Seamless Tube	EN 10297-2	Chemical Compositions
		EN 10297-2	Mechanical Properties
		EN ISO 1127	Tolerances. However, some tolerances are also covered within EN 10297-2 itself
		EN 10079	Definition of product forms
		EN 10002	Tensile test methods
		EN 10246-2	Eddy current testing for leak tightness
		EN 10246-3	Eddy current testing for detection of imperfections
		EN 10246-5	Peripheral MPI testing for detection of longitudinal imperfections
		EN 10246-7	Peripheral ultrasonic testing for detection of longitudinal imperfections
		EN ISO 377	Test piece sampling
		EN ISO 3651-2	Intergranular corrosion testing

Chapter 22 - How Much Do You Know About Stainless Steel?

Hopefully, if you've got this far you will have picked up a lot of information about stainless steel. Here's a short quiz to see how much more you know. The answers can be found on page 113.

1. What element is the main one to make stainless steel corrosion resistant?

   a) Nickel
   b) Silicon
   c) Chromium

2. What is the minimum quantity of this element which defines stainless steel in the European standard?

   a) 11.5%
   b) 10.5%
   c) 12.5%

3. The passive layer on stainless steel is a thin film of:

   a) Chromium Sulphide
   b) Chromium Nitride
   c) Chromium Oxide

4. What is the main element which differentiates 1.4301 (304) from 1.4401 (316)?

   a) Manganese
   b) Nickel
   c) Molybdenum

5. Some stainless steels have low magnetic permeability due to their being mainly:

   a) Austenitic
   b) Ferrite
   c) Martensitic

6. The nickel is used in stainless steels to make them:

   a) More weldable
   b) Improve formability
   c) Both of these

7. Surface finish is important in determining the corrosion resistance of stainless steel:

   a) True
   b) False

8. The surface finish called 2K restricts the surface roughness to a value of Ra = ???

   a) 0.05 micron
   b) 0.5 micron
   c) 5 micron

9. 1.4301 (304) differs from 1.4307 (304L) mainly in the content of which element?

   a) Silicon
   b) Chromium
   c) Carbon

10. Increased nickel content in austenitic stainless steels improves which property related to formability:

    a) Stretch forming
    b) Deep drawing
    c) Both of these

11. Duplex stainless steels are so called because they contain 2 phases:

    a) Austenite and martensite
    b) Austenite and ferrite
    c) Ferrite and martensite

12. Duplex steels are stronger than:

    a) Austenitic stainless steels
    b) Ferritic stainless steels
    c) Both of these

13. Duplex steels generally have:

a) Lower nickel content than austenitic steels
b) Higher nickel content than austenitic steels
c) About the same nickel content

14. The 200 series of steels uses which element in place of nickel:

a) Magnesium
b) Molybdenum
c) Manganese

15. The resistance to pitting corrosion of stainless steels is improved by 3 elements:

a) Chromium, silicon and nitrogen
b) Nickel, manganese and nitrogen
c) Chromium, molybdenum and nitrogen

16. The type of corrosion involving contact between dissimilar metals is called:

a) Intergranular corrosion
b) Galvanic corrosion
c) Stress corrosion cracking

17. The type of stainless steel which is most suitable for very low temperatures is:

a) Ferritic
b) Duplex
c) Austenitic
d) Martensitic

18. The main property of stainless steel required for use at low temperature is:

a) High strength
b) High toughness
c) High thermal conductivity.

19. The corrosion resistance of 1.4401 (316) stainless steel to 20% Sulphuric acid at 20 deg C:

    a) Is similar to that of 1.4301 (304)
    b) Is better than 1.4301 (304)
    c) Is worse than 1.4301 (304)

20. Cold working of austenitic stainless steel can increase the magnetic permeability by the formation of:

    a) A martensitic structure
    b) A ferritic structure
    c) A duplex structure.

21. Which of the following types responds to a heat treatment known as quenching and tempering:

    a) Austenitic
    b) Martensitic
    c) Ferritic
    d) Duplex

22. Stress corrosion cracking is a form of corrosion requiring some kind of stress. The type of stress is:

    a) Compressive stress
    b) Tensile stress
    c) Both of these

23. Sulphur is added to stainless steel to improve:

    a) Corrosion resistance
    b) Formability
    c) Machinability
    d) Thermal expansion

24. Intergranular corrosion is a type of corrosion linked to the formation of:

    a) Chromium chloride
    b) Molybdenum carbide
    c) Chromium carbide

25. Brinell, Vickers and Rockwell are all types of:

a) Hardness test
b) Tensile test
c) Impact test

26. Chromium, silicon, aluminium and nickel are all used to improve the properties of stainless steel for use at:

a) High temperature
b) Low temperature
c) Normal temperature

27. General corrosion is a uniform type of corrosion caused by the attack of the passive film by:

a) Chloride
b) Acids
c) Alkalis

28. In steels, the term "brittle" means having a low value of:

a) Strength
b) Hardness
c) Toughness

29. In addition to purely technical factors the selection of a stainless steel grade may depend on:

a) Cost
b) Availability
c) Risk
d) All of these

30. The production of stainless steel was significantly reduced in cost with the introduction of:

a) The AOD Vessel
b) Continuous casting
c) Both of these

Glossary of Terms Relevant to Stainless Steel

Items in bold are Terms in the Glossary

Term	Explanation
200 Series	A group of **austenitic** stainless steels derived from the AISI numbering system. Based on substituting **manganese** for **nickel** to provide the austenitic structure.
300 Series	A group of **austenitic** stainless steels derived from the AISI numbering system. Based on the use of **nickel** to provide the austenitic structure.
400 Series	A group of **ferritic** and **martensitic** stainless steels derived from the AISI numbering system. Characterised by no or low **nickel.**
Acid Corrosion	Also called **general corrosion.** Characterised by uniform metal loss. Contrast with **pitting corrosion.**
Alloy	A combination of two or more **metals** or other **elements. Steel** is an **alloy** of **iron** and **carbon.**
Alumina	Aluminium oxide (Al_2O_3). An abrasive used on polishing belts and wheels to give a directional polish. On a micro-level gives a non-uniform appearance. Prone to trapping contaminants and giving disappointing results in external architectural applications.
Aluminium	An **element** (symbol Al) used to improve **oxidation** resistance in **ferritic** stainless steels and strength in **precipitation hardening** steels.
Annealing	A general term in **heat treatment** usually referring to a softening process.
AOD Vessel	Argon oxygen decarburisation vessel. A development in the refining process allowing cheaper production of stainless steel.
Atom	The smallest particle of an **element.**
Austenitic	The most common type of stainless steels. The austenitic structure gives the characteristic properties of this type including: **Formability** **Weldability** **Work hardening** **Non-magnetic** At an atomic level austenitic is a face centred cubic structure.

Billet	The starting point for the production of long products such as **rod.** May be **continuously cast** or rolled from **ingot or bloom.** Typical section size 150mm.
Bloom	The starting point for the production of long products such as **bar** and **rod.** May be **continuously cast** or rolled from **ingot.** Typical section size 300-400 mm.
Brearley	Harry Brearley is the Sheffield metallurgist connected with the invention of stainless steel.
Bright annealing	The softening of stainless steel in an inert atmosphere to preserve the bright cold rolled surface.
Brittle	Showing low level of energy absorbed in an **impact toughness** test.
Carbon	The essential **element** (symbol C) added to **iron** to make **steel.** In stainless steel, the range of **carbon** content is very wide from about 0.015 – 1% depending on the properties required.
Casting	A product that is used in the as-cast condition, that is without mechanical working. The main benefit is that intricate shapes can be formed by pouring liquid metal into a mould. Castings have poorer mechanical properties compared to wrought products due to their coarser microstructure.
Chemical symbol	An abbreviation of the name of an element. **Chromium** = Cr.
Chloride	A type of **ion**, symbol Cl^-, which is among the most common chemical species to be detrimental to the passive film on stainless steel. Seawater contains about 3% NaCl (sodium chloride). Chloride can cause **pitting corrosion, crevice corrosion** and **stress corrosion cracking.**
Chromium	The metallic **element** (symbol Cr) which is in all stainless steels. Forms a passive oxide layer which prevents **corrosion.**
Compound	A combination of two or more **elements** which is quite different in nature to any of the constituents.
Continuous casting	A method of casting allowing continuous casting of **slab, bloom** or **billet.** Reduced cost and improved yield compared to individual **ingot** casting.

Continuous casting	A method of casting allowing continuous casting of **slab, bloom** or **billet.** Reduced cost and improved yield compared to individual **ingot** casting.
Continuously Produced Plate (CPP)	Plate cut from hot rolled coil. A maximum thickness of about 13 mm in widths of up to 2000 mm can be produced via this process route. Sometimes called coil plate.
Copper	**Element** (symbol Cu) normally found in all steels in small quantities. Deliberately added to stainless steel to improve resistance to acids such as sulphuric acid and to lower the work hardening for cold heading of fasteners.
Corrosion	The attack of a **metal** or **alloy** by a chemical substance. It is electrochemical in nature.
Creep	A type of deformation which depends on the time as well as the **stress** applied. In steels this phenomenon is important above about 550°C.
Crevice Corrosion	A type of **corrosion** resulting from the exclusion of oxygen from between tightly contacting surfaces.
Critical Crevice Temperature (CPT)	The temperature at which a stainless steel starts to show **crevice corrosion** in a standard laboratory solution. Used for comparing stainless steels.
Critical Pitting Temperature (CPT)	The temperature at which a stainless steel starts to show **pitting corrosion** in a standard laboratory solution. Used for comparing stainless steels.
Crystal	A structure which, at the atomic level, has a regular arrangement of **atoms.**
Deep Drawing	A method of forming involving pressing a sheet into a hollow mould without restraining the outer edges of the sheet. Compare with **stretch forming.**
Ductile	Two distinct meanings: 1) A material is said to be ductile if it shows a high level of elongation in the **tensile** test. 2) A material is said to be ductile if it shows a high level of absorbed energy in an **impact toughness** test.
Ductile/Brittle Transition Temperature	The temperature at which a series of impact test specimens shows 50% **brittle** and 50% **ductile** fracture surfaces.

Duplex	A type of stainless steel having approximately 50% **austenite** and 50% **ferrite.** This gives it higher strength than either structure on its own.
Elongation	In a **tensile test,** the % increase in the gauge length on the test sample.
En	"Emergency number". An obsolete (from 1970) type of grade used in British standards. It is still used mainly because many drawings and specifications in use date back before the official date of obsolescence.
EN Standard	European Norm. EN Standards are published in the main European languages by each national standards body. BS EN is published by British Standards in English, DIN EN in German and AFNOR EN in French.
Ferritic	A type of stainless steel based on **chromium** and only small additions of other **elements.** At an atomic level ferritic is a body centred cubic structure.
Forging	The deformation of metal usually from high temperature. Open-die forging allows a rough approximation of the final shape to be achieved. Closed-die forging or drop forging using a mould allows a shape much closer to the final shape to be achieved.
Formability	Loose term covering a wide range of processes. Generally it means the ability to be formed into complex shapes. Formability is dependent on the grade of steel, its **mechanical properties** and the forming operation. It is important to note that formability in one type of operation may not carry over to another. **Stretch forming** and **deep drawing** are examples of forming operations.
Formula	Shorthand way of writing a **compound.** Example Fe_2O_3 is **iron** oxide.
Galvanic corrosion	A type of **corrosion** involving the contact of dissimilar metals joined by an electrolyte. In some circumstances stainless steel and aluminium in contact can accelerate the corrosion of the aluminium.
Grade	A type of **steel** or other metallic **alloy** with defined chemical composition limits.
Grit size	The nominal size of abrasive particles on a polishing belt or wheel. 180 240 and 320 are common grit sizes. The abrasive material e.g. **alumina** or **silicon carbide** is important in determining the surface roughness.

Hardness	The ability of a material to withstand the indentation by a hardened steel or diamond indenter. Methods of hardness testing include Brinell, Vickers and Rockwell. Charts showing conversions between different methods and conversion to **UTS** should be treated with caution.
Heat treatment	The use of heating and cooling of a substance usually with the intention of modifying its **microstructure** and therefore its **mechanical properties.** Heat treatments relevant to stainless steel include: **Solution annealing** **Quenching** **Tempering** **Stress relieving** **Precipitation hardening**
Ion	An **atom** or **molecule** which has an excess or lack of electrons thereby giving it a positive or negative electric charge.
Impact toughness	The ability of a material to resist a sudden impact. Measured in the Charpy test by allowing a swinging hammer to hit a small notched sample and measuring the distance swung after the impact. All steels except austenitic types show a sudden loss of impact toughness at low temperatures.
Ingot	A discrete lump of cast metal poured into individual moulds. Depending on shape can be further rolled into **slab** or **bloom.** Also used in **forging.**
Intergranular corrosion	A form of **corrosion** caused by the formation of chromium carbide which reduces the **chromium** content of the steel below that required to form the **passive film.**
Iron	Metallic **element** (symbol Fe) which is the basis for all **steels.** In its pure form it is soft.
Lattice	A pattern or regular arrangement of **atoms** often in a simple geometric shaped such as a cube or prism.
Lean Duplex	An imprecise term applied to **duplex** stainless steels with a lower alloy content than 2205 (1.4462) grade duplex steel.
Magnetic Permeability	More accurately Relative Magnetic Permeability. A measure of a material's ability to be attracted by a magnet. A relative magnetic permeability of 1 means completely non-magnetic.

Manganese	**Element** (symbol Mn) normally found in all steels. Added to stainless steel as an alternative to **nickel** to give the **austenitic** structure in the **200 series.**
Martensitic	A type of stainless steel based on **chromium** and small additions of other elements. Levels of **carbon** can be high, allowing the quenching and tempering of the steel to very high strength.
Molecule	The smallest particle of a **compound.** A molecule contains 2 or more **atoms.**
Metal	An **element** that is usually shiny, easy to form, conducts heat and electricity well.
Molybdenum	Metallic **element** (symbol Mo) used to improve resistance to **pitting** and **crevice corrosion** in stainless steels.
Multiple Certification	The practice of certifying a batch of steel to more than one grade or standard. Allows more efficient production in the melting shop and more flexibility at the stockholder. Common examples are: 1.4301/1.4307 (304/304L) 1.4401/1.4404 (316/316L) EN 10088-2/EN 10028-7 EN 10088-2/ASTM A240 EN 10088-3/BS 970
Nanometre	One millionth of a millimetre or 10^{-9} metres. The **passive film** on stainless steel is a few nanometres thick.
Nickel	Metallic **element** (symbol Ni) used in stainless steel to give improved **weldability** and **formability.** Also improves high temperature **oxidation** resistance.
Niobium	A metallic element (symbol Nb) used in stainless steels to prevent formation of chromium carbide which can lead in turn to **intergranular corrosion.**
Nitrogen	An **element** (symbol N) which is used to give higher strength, increased **pitting corrosion** resistance and lower **magnetic permeability** in stainless steels.
Oxidation	The high temperature combination of a **metal** or **alloy** with oxygen in the atmosphere.
Passivation	The process of forming the **passive film** on the surface of stainless steel. Stainless steel passivates in normal atmospheric conditions. However, passivation can be accelerated using acids such as nitric or citric.

Passive Film	The **chromium** oxide layer that forms on stainless steel to give it **corrosion** resistance.
Pickling	The removal of high temperature oxide from the surface of stainless steel. High temperature oxide is formed during processes such as hot working and welding. If left un-pickled high temperature oxide reduces corrosion resistance. Strong acids such as a mixture of nitric and hydrofluoric are commonly used.
Pitting Corrosion	**Corrosion** characterised by local attack. Caused by chemical species notably **chloride ions.** Contrast with general corrosion.
Pitting Resistance Equivalent Number (PREN)	A measure of the ability of a stainless steel to resist **pitting corrosion.** Calculated from the formula: PREN = %Cr + 3.3 x (%Mo + 0.5 x %W) + 16 x % N
Polished	An overall term covering a wide range of surface finishes on stainless steel including **dull polished, satin polished, bright polished.** In defining a polished finish it is advisable to use **surface roughness, Ra,** to describe the required finish.
Precipitation Hardening	A heat treatment in certain special steels which forms tiny particles leading to high strength.
Proof Stress	In a **tensile test,** stress at which the sample shows a particular strain, often 0.2%. Used in place of a defined yield stress.
Quarto Plate	Hot rolled plates made from rolled **slab** on a reversing mill. Quarto plates remain flat throughout processing. Thicknesses of up to 150 mm can be produced by this method. Contrast with **Continuously Produced Plate.**
Ra	A measure of surface roughness. An important factor in determining the **corrosion** resistance of a stainless steel surface. Measured in micron = one thousandth of a millimetre. An Ra of < 0.5 micron is regarded as acceptable for external architectural applications.
Residual Element	An element which is not deliberately added to stainless steel. These elements cannot be removed during the steelmaking process. Copper, tin, zinc and lead are examples.
Rust	**Iron** oxide. The product of corrosion in steels.

Sensitisation	The precipitation of **chromium carbide** as a result of holding stainless steel at temperatures around 650°C. Can give rise to **intergranular corrosion.** Prevented by using low **carbon** < 0.030% or using **titanium** or **niobium** to combine preferentially with **carbon.**
Silicon	**Element** (symbol Si) normally found in all steels. Added to stainless steel for high temperature oxidation resistance.
Silicon carbide	SiC. An abrasive used on polishing belts and wheels to give a directional polish. On a micro-level gives a uniform appearance. For the same grit size gives a much better **corrosion** performance than **alumina.**
Steel	An **alloy** of **iron** and **carbon.** Other elements added to steel to improve the mechanical properties include **chromium, nickel, molybdenum** and **manganese.**
Steel Name	In the EN Standards, this is a long description of a **grade** of steel. For example, X5CrNi18-10. This is the Steel Name for Steel Number 1.4301. The X = stainless. Cr and Ni show the main **elements** and the 18 10 show the approximate contents of the **elements.**
Steel Number	In the EN Standards, this is the short description of a **grade** of steel. For example, 1.4301. The number is partly meaningful. The first "1" indicates a steel. "4" indicates a stainless steel. The last 3 digits are effectively arbitrary.
Strain	The increase per unit length when a material has a **stress** applied to it.
Stress	The load applied to a material divided by the area over which it is applied. Common units of stress include: N/mm^2 = Newtons per square millimetre MPa = Megapascals (identical to N/mm2) These are used in EN Standards tsi = tons per square inch This was used in the old BS standards psi = pounds per square inch ksi = thousands of pounds per square inch Used in American standards

Stress Corrosion Cracking	A type of **corrosion** requiring a tensile **stress,** a sufficiently high temperature and a corrosive environment.
Stretch Forming	A method of forming involving pressing a sheet into a hollow mould whilst restraining the outer edges of the sheet allowing stretching to occur. Compare with **deep drawing.**
Sulphur	An **element** (symbol S) used to improve the machinability of steels. In genuine free-machining steels a content of at least 0.15% S is used.
Superduplex	A generic term applied to **duplex** stainless steels with at least 25% **chromium.**
Tempering	A **heat treatment** usually following a hardening operation which softens **martensitic** steels to provide a useful combination of strength and **ductility.**
Temper Rolling	A rather misleading term meaning cold working to produce an increase in strength. Stainless steel can be supplied in a number of "tempers" e.g. quarter hard, half hard, three quarter hard, full hard. These correspond to increasing levels of strength/hardness and decreasing elongation and therefore formability.
Tensile Test	A mechanical test involving the slow pulling apart of a standard test sample. The test measures: 0.2% **Proof Stress** **Ultimate Tensile Stress (UTS)** **Elongation**
Thermal Conductivity	The ability of a material to allow heat to pass through it. **Austenitic** stainless steels have a lower value than **ferritic** stainless steels or **carbon** steels.
Thermal Expansion Coefficient	A measure of the increase in size of a material with increased temperature. **Austenitic** stainless steels have a higher value than **ferritic** stainless steels or **carbon** steels.
Titanium	A metallic element (symbol Ti) used in stainless steels to prevent formation of **chromium** carbide which can lead in turn to **intergranular corrosion.**
Tungsten	A metallic **element** (symbol W) used to improve **pitting corrosion** in some **duplex** stainless steels.
Ultimate Tensile Stress (UTS)	In a **tensile test,** the maximum stress measured on the sample before the sample fractures.

Weldability	A general term meaning that a material can be welded without requiring pre-heating or post-heating, tight control of welding parameters and resulting in welds with good mechanical properties. **Austenitic** stainless steels are regarded as having good weldability as they can be welded in a wide range of section thicknesses. **Ferritic** steels are less so as they can only be welded in thin sections. **Duplex** steels can be welded with care required in selection of welding consumables and control of welding parameters. **Martensitic** steels are not easily welded and often need pre-heating and post-weld heat treatment. Low carbon martensitic steels have been developed to avoid pre and or post heating.
Weld Decay	Equivalent to **intergranular corrosion.** So called because early high **carbon** stainless steels were found to corrode near welds due to **chromium carbide** precipitation.
Work Hardening	The property of inducing increased strength to a material as it is being deformed. Also called cold working or **temper rolling. Austenitic** stainless steels have a high degree of work hardening.
Young's Modulus	The ratio of **stress** to **strain** on the straight line portion of a stress-strain curve in a tensile test. Does not change much with grade.

Answers to Stainless Steel Quiz

1.c) 2.b) 3.c) 4.c) 5.a) 6.c) 7.a) 8.b) 9.c) 10.b)

11.a) 12.c) 13.a) 14.c) 15.c) 16.b) 17.c) 18.b) 19.b) 20.a)

21.b) 22.b) 23.c) 24.c) 25.a) 26.a) 27.b) 28.c) 29.d) 30.c)

Index

Sources of Information on Stainless Steel

BSSA – www.bssa.org.uk

Euro Inox – www.euro-inox.org

International Stainless Steel Forum (ISSF) - www.worldstainless.org

International Chromium Development Association (ICDA) - www.icdachromium.com

Nickel Institute - www.nickelinstitute.org

International Molybdenum Association (IMOA) - www.imoa.org.uk

Steel Construction Institute (SCI) - www.steel-sci.org

National Stainless Steel Associations - www.bssa.org.uk/organisations.php

Stainless Steel Producers' Websites

- To be part of the BSSA network of companies that represent all parts of the stainless steel supply chain and to contribute your ideas for the development of the market.

- To have the opportunity to participate in a range of committees and working parties including the Council, Marketing & Technical Committee, Industry Forum, Finishing Section.

- To have your company's full details listed on our very popular website and for your products and services to appear in the Find A Supplier section; Members who deal in special grades can be included in the very popular article Special Grades of Stainless Steel - Where to Find Them.

- To get sales leads; the BSSA receives hundreds of enquiries each year that are related to sourcing of products and services – member companies are always recommended first.

- Access to Independent Buyers Ltd, a leading procurement support business to run a purchasing consortium on behalf of our members. They are able to offer advice and support on all areas of procurement, and give members access to their existing purchase agreements.

- Benefit from links to Euro Inox, the European Stainless Steel Development Association, the International Stainless Steel Forum and other market development organisations across the world.

- Generous discounts for events, training and publications.

- Access to the members' area of the website.

Membership Enquiry Form

For further information about the BSSA please complete this form and fax it to 0114 266 1252 or:
Send and email to enquiry@bssa.org.uk or
Go to www.bssa.org.uk/contact_membership.php and fill in the on-line form

Name of Company:

Address:

Tel No:

Fax No: Email:

Website:

Main Contact:

Job Title: